Philipp Limbourg

Dependability Modelling under Uncertainty

Studies in Computational Intelligence, Volume 148

Editor-in-Chief
Prof. Janusz Kacprzyk
Systems Research Institute
Polish Academy of Sciences
ul. Newelska 6
01-447 Warsaw
Poland
E-mail: kacprzyk@ibspan.waw.pl

Further volumes of this series can be found on our homepage: springer.com

Vol. 126. Oleg Okun and Giorgio Valentini (Eds.)
Supervised and Unsupervised Ensemble Methods and their Applications, 2008
ISBN 978-3-540-78980-2

Vol. 127. Régie Gras, Einoshin Suzuki, Fabrice Guillet and Filippo Spagnolo (Eds.)
Statistical Implicative Analysis, 2008
ISBN 978-3-540-78982-6

Vol. 128. Fatos Xhafa and Ajith Abraham (Eds.)
Metaheuristics for Scheduling in Industrial and Manufacturing Applications, 2008
ISBN 978-3-540-78984-0

Vol. 129. Natalio Krasnogor, Giuseppe Nicosia, Mario Pavone and David Pelta (Eds.)
Nature Inspired Cooperative Strategies for Optimization (NICSO 2007), 2008
ISBN 978-3-540-78986-4

Vol. 130. Richi Nayak, Nikhil Ichalkaranje and Lakhmi C. Jain (Eds.)
Evolution of the Web in Artificial Intelligence Environments, 2008
ISBN 978-3-540-79139-3

Vol. 131. Roger Lee and Haeng-Kon Kim (Eds.)
Computer and Information Science, 2008
ISBN 978-3-540-79186-7

Vol. 132. Danil Prokhorov (Ed.)
Computational Intelligence in Automotive Applications, 2008
ISBN 978-3-540-79256-7

Vol. 133. Manuel Graña and Richard J. Duro (Eds.)
Computational Intelligence for Remote Sensing, 2008
ISBN 978-3-540-79352-6

Vol. 134. Ngoc Thanh Nguyen and Radoslaw Katarzyniak (Eds.)
New Challenges in Applied Intelligence Technologies, 2008
ISBN 978-3-540-79354-0

Vol. 135. Hsinchun Chen and Christopher C. Yang (Eds.)
Intelligence and Security Informatics, 2008
ISBN 978-3-540-69207-2

Vol. 136. Carlos Cotta, Marc Sevaux and Kenneth Sörensen (Eds.)
Adaptive and Multilevel Metaheuristics, 2008
ISBN 978-3-540-79437-0

Vol. 137. Lakhmi C. Jain, Mika Sato-Ilic, Maria Virvou, George A. Tsihrintzis, Valentina Emilia Balas and Canicious Abeynayake (Eds.)
Computational Intelligence Paradigms, 2008
ISBN 978-3-540-79473-8

Vol. 138. Bruno Apolloni, Witold Pedrycz, Simone Bassis and Dario Malchiodi
The Puzzle of Granular Computing, 2008
ISBN 978-3-540-79863-7

Vol. 139. Jan Drugowitsch
Design and Analysis of Learning Classifier Systems, 2008
ISBN 978-3-540-79865-1

Vol. 140. Nadia Magnenat-Thalmann, Lakhmi C. Jain and N. Ichalkaranje (Eds.)
New Advances in Virtual Humans, 2008
ISBN 978-3-540-79867-5

Vol. 141. Christa Sommerer, Lakhmi C. Jain and Laurent Mignonneau (Eds.)
The Art and Science of Interface and Interaction Design (Vol. 1), 2008
ISBN 978-3-540-79869-9

Vol. 142. George A. Tsihrintzis, Maria Virvou, Robert J. Howlett and Lakhmi C. Jain (Eds.)
New Directions in Intelligent Interactive Multimedia, 2008
ISBN 978-3-540-68126-7

Vol. 143. Uday K. Chakraborty (Ed.)
Advances in Differential Evolution, 2008
ISBN 978-3-540-68827-3

Vol. 144. Andreas Fink and Franz Rothlauf (Eds.)
Advances in Computational Intelligence in Transport, Logistics, and Supply Chain Management, 2008
ISBN 978-3-540-69024-5

Vol. 145. Mikhail Ju. Moshkov, Marcin Piliszczuk and Beata Zielosko
Partial Covers, Reducts and Decision Rules in Rough Sets, 2008
ISBN 978-3-540-69027-6

Vol. 146. Fatos Xhafa and Ajith Abraham (Eds.)
Metaheuristics for Scheduling in Distributed Computing Environments, 2008
ISBN 978-3-540-69260-7

Vol. 147. Oliver Kramer
Self-Adaptive Heuristics for Evolutionary Computation, 2008
ISBN 978-3-540-69280-5

Vol. 148. Philipp Limbourg
Dependability Modelling under Uncertainty, 2008
ISBN 978-3-540-69286-7

Philipp Limbourg

Dependability Modelling under Uncertainty

An Imprecise Probabilistic Approach

Dipl.-Inform. Philipp Limbourg
University of Duisburg-Essen
Department of Engineering/Information Logistics
Bismarckstr. 90 (BC 509)
47057 Duisburg
Germany
E-Mail: p.limbourg@uni-due.de

A dissertation approved by the faculty of engineering, University of Duisburg-Essen (Germany) for the degree of Doctor of Engineering.

Examination date: April 16, 2008

Reviewers:

Prof. Dr. Hans-Dieter Kochs, examiner
Prof. Dr. Fevzi Belli, co-examiner

ISBN 978-3-642-08880-3 e-ISBN 978-3-540-69287-4

DOI 10.1007/978-3-540-69287-4

Studies in Computational Intelligence ISSN 1860949X

Printed in acid-free paper

9 8 7 6 5 4 3 2 1

springer.com

To
Maria, Kurt, Nora, Vicky and Marina
for guiding me through a world full of uncertainties.

Preface

Mechatronic design processes have become shorter and more parallelized, induced by growing time-to-market pressure. Methods that enable quantitative analysis in early design stages are required, should dependability analyses aim to influence the design. Due to the limited amount of data in this phase, the level of uncertainty is high and explicit modeling of these uncertainties becomes necessary.

This work introduces new uncertainty-preserving dependability methods for early design stages. These include the propagation of uncertainty through dependability models, the activation of data from similar components for analyses and the integration of uncertain dependability predictions into an optimization framework. It is shown that Dempster-Shafer theory can be an alternative to probability theory in early design stage dependability predictions. Expert estimates can be represented, input uncertainty is propagated through the system and prediction uncertainty can be measured and interpreted. The resulting coherent methodology can be applied to represent the uncertainty in dependability models.

To enhance the base data in early design stages, a new way of reusing data from previous dependability analyses is developed. Neural networks and Gaussian processes (Kriging) capture similarity relations with in-service components to predict the new component's probability of failure. In a case study, failure distributions of mechatronic components are predicted with acceptable accuracy. To support early design decisions, design-screening, discovering a whole subset of optimal candidate systems in a single optimization run, is realized by using feature models and multi-objective evolutionary algorithms. With this specification method, it is possible to represent arbitrary redundancy allocation problems (RAP) which are commonly analyzed in literature, but also more complex and especially more realistic design spaces. An evolutionary algorithm for uncertainty objectives is designed to screen this set for optimal systems.

The proposed methods are applied to estimate the compliance of an automatic transmission system to an IEC 61508 SIL. Conformance to a failure probability threshold is deduced under high uncertainty. A sensitivity analysis on the nonspecifity identifies components with a high contribution to the overall uncertainty. By means of feature models, a design space with several different component and subsystem structures is defined and successfully screened, resulting in a set of systems with good dependability-cost trade-offs.

I would like to take this opportunity to extend my most heartfelt gratitude and words of appreciation to those, who aided and fostered the accomplishment of this research work.

My sincere thanks go to Prof. Dr. Kochs and Dr. Jörg Petersen for all the guidance and encouragement they offered me throughout the research period. Without their inspiration and support, I would never have completed this thesis on time. I would also like to thank Prof. Dr. Fevzi Belli, who agreed to be my second reader.

There are a number of people, whose collaboration enriched this work: To Dr. Robert Savić of ZF Friedrichshafen AG, who provided me with material for a case study and who was also an extraordinary industrial advisor, I say thank you. I'd also like to thank Katharina Funcke for being a great beta reader. My sincere gratitude goes to the members of the ESReDA workgroup on uncertainty, especially Fabien Mangeant of EADS and Etienne de Rocquigny of EdF R&D for the fruitful discussions and new insights into an uncertain world.

My gratitude is extended to my colleagues and friends at the chair of Information Logistics: Ingo Dürrbaum, Michael Gallinat, Thomas Hensel, Daniela Lücke-Janssen and Kirsten Simons, who accompanied me throughout the long and winding journey until the completion of this work and Katja Kehl and Thomas Günther, who shielded us from everyday troubles during this journey.

Last but not the least, I am deeply grateful to my family: Maria, Kurt, Nora, Vicky and to my girlfriend Marina. They gave me their entire and much needed loving care and emotional support, shared their inspiring experiences and were always there when I needed them. For these reasons and many more, I dedicate this work to you.

Duisburg, Germany
April 2008

Philipp Limbourg

Contents

Acronyms

α	Weibull characteristic lifetime
β	Weibull shape parameter
$\widetilde{X}^i$	Belief vector pinched at index i
$\check{F}(...)$	Cumulative mass function
$\check{\xi}_1, ...\check{\xi}_n$	Predicted outputs
$\check{F}_X(...), \overline{F_X}(...), \underline{F_X}(...)$	Best estimate, upper and lower bound of a predicted function F_X
ε	Random noise factor
$\underline{\overline{g}} = [\underline{g}, \overline{g}]$	Multidimensional focal element
$\underline{\overline{g}} = [\underline{g}, \overline{g}] \in G$	Focal element, set of focal elements
κ, k	GP covariance matrix parts
$\lambda_{Weib}(t)$	Weibull failure rate
$\mathscr{P}(X)$	Power set of X
$\mu(...), \sigma(...)^2$	Mean and variance functions of a Gaussian process
$\Psi = (\psi_1, ..., \psi_n)$	Training inputs
$\Xi = (\xi_1, ..., \xi_n)$	Training outputs
Cov_n	Covariance matrix of a n-dimensional Gaussian distribution
$c_s \in C_s$	Component state vector, set of component state vectors
d	Fixed parameter vector
p	Component failure probability vector
$q = (q_{dist}, q_{qual}, q_{wo})^T$	Qualifier vector with distance, quality and wear out qualifiers
$w \in W$	Weight vector
$\widehat{f}(...)$	Imprecise probabilistic function
$\widehat{f}(...)$	Probabilistic function
ϕ	System structure function
$\succ_P$	Pareto dominance
$\succ_\alpha$	Alpha-dominance
$\succ_{DS}$	Imprecise probabilistic dominance
$\succ_{Pr}$	Probabilistic dominance
φ	Feature set
$\varphi^+ \in \Phi^+$	Feature set with more than one realization
ϖ	Transformation factor

$AW(...)$ Aggregated width measure
$Bel(...), Pl(...)$ Belief / Plausibility of an event
$C(...)$ Copula function
$C_+(...), C_-(...), C_P(...)$ Fréchet bound copulas, product copula
C_A, F_A Component C_A with known failure function F_A used for similarity prediction
$C_G(...), \rho$ Gaussian copula with copula correlation matrix ρ
C_X, F_X Unknown component C_X with unknown failure function F_X
$d(...)$ Density measure
$d\angle_k(...)$ K-nearest angle
$d_k(...)$ K-nearest neighbor
$DN(...)$ Dissonance measure
$E(...)$ Expected value
$e(...)$ Neural network error criterion
el Elitism number
$Err(...)$ Error function
F Cumulative distribution function (CDF)
$F_{Bet}(...)$ Beta distribution
$f_{Cov}(...)$ Covariance function of a Gaussian process
$F_{Gauss}(...), F^{n,\rho}_{Gauss}(...)$ Normal distribution, multivariate distribution with correlation matrix ρ
$f_{NN}(...)$ Neural network regression function
$f_{Pri}(...)$ Transfer factor prior probability distribution
$f_{reg}(...)$ Regression function
$f_{sim}(...)$ Similarity prediction function
$F_{Uni}(...)$ Uniform distribution
$F_{Weib}(...)$ Weibull distribution
$Fr(...)$ Nondominated front number
$GH(...)$ Generalized Hartley measure
$H(...)$ Shannon entropy
I Gaussian process input dimension
K Conflict between focal elements
$m(...)$ Basic probability assignment
$m_{1,...,n}(...)$ Joint BPA
Mat Mating pool
$Med(...)$ Median
$P(...)$ Probability of an event
P_c Crossover probability
P_m Mutation probability
p_s System failure probability
Pop Population
$Q(...)$ Quantile function
$Q_5(...), Q_{95}(...)$ Quantile
$R(...)$ Reliability
$Ran(f)$ Range of function f
$rsize$ Repository size

$s \in S_s$	System state, Set of system states
$s_{AW,i}$	Sensitivity indices on the *AW* measure
$s_{GH,i}$	Sensitivity indices on the *GH* measure
Z_n	GP normalization constant
$x \in X$	Deterministic variable
x	Vector (bold type)
$\overline{\underline{x}} = [\underline{x}, \overline{x}]$	Interval, lower and upper bound
$\widehat{X}$	Random variable
$\widetilde{X}$	Belief variable
$\widehat{X}$	Joint random variable
$\widetilde{X}$	Joint belief variable
ADF	Average discretization function
ATM	Automatic transmission
BPA	Basic probability assignment
CAN	Controller area network
CDF	Cumulative distribution function
DFS	Depth-first search
DST	Dempster-Shafer theory of evidence
EDC	Electronic Diesel control
EDS	Early design stage
ESReDA	European safety and reliability data association
FMEA	Failure mode and effects analysis
(G, M)	Random set
GP	Gaussian process learning
MOEA	Multi-objective evolutionary algorithm
MTTF	Mean time to failure
NN	Neural network
NSGA	Non-dominated sorting genetic algorithm
PDF	Probability density function
RAP	Redundancy allocation problem
RBD	Reliability block diagram
SIL	Safety integrity level
SIPTA	Society of Imprecise Probabilities and Their Applications
SPEA	Strength Pareto evolutionary algorithm
VDA	Verband der Automobilindustrie

1 Introduction

There has been tremendous progress in mechatronic engineering in the last decade. With the ongoing integration of the different domains united by the keyword "Mechatronics", systems growing both in functionality and complexity have been developed. In mechatronics, projects can subsume whole product families, thus it is no longer the single product but the product family that is designed and validated using reliability and safety methods. The project cycles have become shorter and more parallelized, induced by ever growing time-to-market pressure. Development processes have become highly structured and formalized and the applied process models such as the V-Model require simulation and modeling tools that accompany the development process from its very beginning. Otherwise, the project runs the danger of iterative redesigns and along with this, an increase of the project risk.

Therefore, for efficiently contributing to the design of new systems, the integration of a dependability method into the entire development process is important. If the results were provided in an early design stage, they could influence design decisions towards a dependable product [83]. However, the first quantitative reliability and safety analyses are still mainly done at the final stages of the development cycle, too late for a real feedback influence. Dependability predictions consequently lose their importance as a decision-support tool and solely become validation methods for the end product. Many companies suffer from increasing dependability problems, as a result of this, as outlined in chapter 2.

To countervail against this trend, it is necessary to develop and adapt methods, which permit quantitative dependability prediction in early design stages (EDS). While most quantitative dependability analysis methods presume exact input data (e. g. exactly known component failure probabilities), dependability prediction in EDS suffers from the absence of quantitative dependability data and accurate models. This is one of the main reasons, why most companies rely on FMEA or other qualitative methods to assess EDS dependability. It is however often overlooked that expert knowledge and data from similar in-service components are information sources with high availability. These types of data, albeit fuzzy in nature, are not inaccessible for quantitative methods that incorporate uncertainty handling. As common modeling techniques such as fault trees and block diagrams have proven their strength in quantitative analyses, it is tempting to use these techniques already in EDS. Results derived from quantitative

P. Limbourg: Dependability Modelling under Uncertainty, SCI 148, pp. 1–5, 2008.
springerlink.com

dependability predictions, even if highly uncertain, can be used for assessing dependability, testing for compliance with reliability and safety targets or comparing design alternatives. In other words, they are much more flexible in their use. Thus, a purely qualitative approach is not optimal because it limits modeling freedom and future incorporation of quantitative data into the prediction.

Quantitative EDS predictions however can only penetrate into industrial practice if they generate an added value compared to conventional, qualitative predictions. This is possible if quantitative results can be obtained without excessively increasing the modeling effort or if the benefit of quantitative predictions incorporating uncertainty analyses is larger (value of information) than the modeling costs. A chance for reducing the modeling overhead is that quite often a project subsumes several variants of the same product and/or a history of preceding products. Cross-combining the dependability analysis of several variants and reusing data and system models may therefore both reduce the modeling effort and increase the accuracy of and trust in a prediction.

Because of the described limits on the input data, dependability analysis in an EDS needs to deal with a lot of uncertainties both in the model parameters and in the model itself. On the other hand, guidelines such as the IEC 61508 [78] allow for different levels of safety according to the confidence in the safety prediction. Hence, for developing an EDS dependability prediction, it is important to respect two characteristics of EDS: First, the level of uncertainty is high. Regardless of the elaborateness of the prediction methods, input parameters such as component failure probabilities remain rough estimates. Second, analysis results are only useful if trust can be put into them. If a decision-maker has no trust in the dependability prediction, it will not influence the further design process and therefore it will be rendered useless. This points lead to important requirements for EDS dependability prediction:

First, confidence in the results cannot be gained if dependability studies in EDS produce exact results from uncertain, vague inputs. A trustable prediction method will return imprecise results comprising uncertainty bounds if the input data is uncertain, too. Second, confidence cannot be gained if experts who estimate the dependability figures cannot express their degree of uncertainty. An applicable method should therefore have a high flexibility of specifying input uncertainties. And third, if analysis outcomes are utilized for further use such as the investigation of compliance with a threshold, at least three outcomes should be possible:

1. The system complies with the threshold.
2. The system violates the threshold.
3. The information is neither sufficient to infer threshold compliance nor threshold violation.

If the term "design-for-dependability" is taken seriously, it may even be convenient to include dependability predictions into a formal optimization framework for finding an optimal, dependable design. In EDS, it is however neither possible nor desired to find the single, optimal design. Reasons, which account for this, are the amount of uncertainty in a prediction that prevents the declaration of one system being definitely optimal and the interest of the decision-maker in a decision support but not in one single selected design. Human engineers would prefer to select the future designs to pursue themselves. Thus a design screening system should aim for proposing a set of systems matching the interests of the engineers.

The characteristics described in this section lead to requirements for successful EDS dependability predictions that can be given in a nutshell as: Provide an approximate dependability prediction with uncertainty estimate that can be introduced in a decision-making framework. Provide it as soon as possible with little modeling effort.

1.1 Thesis Aims

Following the demands given in the previous section, the work concentrates on the development of methods for supporting the integration of quantitative dependability assessment into EDS. These include the propagation of uncertainty through dependability models, the activation of dependability data from similar components for analyses and the integration of uncertain dependability predictions into an optimization framework.

The main focus of this work is on the application of Dempster-Shafer theory for dependability prediction under uncertainty. Dempster-Shafer theory (DST) is a mathematical representation for uncertainty, which can be seen as an alternative to probability theory. Due to its higher flexibility, it has become increasingly popular in the last years. However, applications to system dependability are rather limited and in this thesis it will be examined whether DST can be a serious competitor to probability theory in this area.

Another problem in EDS is the small amount of data. A novel way of reusing data from previous or parallel dependability analyses is developed in order to enhance the data situation. Data from previous projects can be considered as quite accurate, especially if the corresponding system is already fielded. By combining expert estimates on the similarity between old and new systems with the data base of a past project, this method tries to give predictions on the new system's dependability data.

However, it is not only sufficient to predict the dependability of a single design alternative. Especially in an EDS, where the overall design is still unclear, it can be tempting to screen a lot of candidate designs for the ones with an optimal dependability-cost trade-off. To make such a screening method accessible, it is necessary to provide efficient specification methods for a realistic design space. This thesis introduces feature models as a tool, which engineers can use for rapidly and intuitively defining a large design space with complex designs. To aid in the investigation of this design space, multi-objective evolutionary algorithms for uncertain objectives are developed and applied. As a result of the screening process, the decision-maker receives a set of systems, each being optimal for a specific dependability/cost trade-off.

1.2 Overview

This thesis is structured as shown in figure 1.1. In chapter 2, the topic of EDS dependability prediction is introduced. An overview on mechatronic process models such as the V-Model is given and the integration of dependability issues into this process models is discussed. Characteristics of EDS dependability predictions are outlined and definitions and fundamentals of dependability-related concepts are given. Chapter 3 comprises of

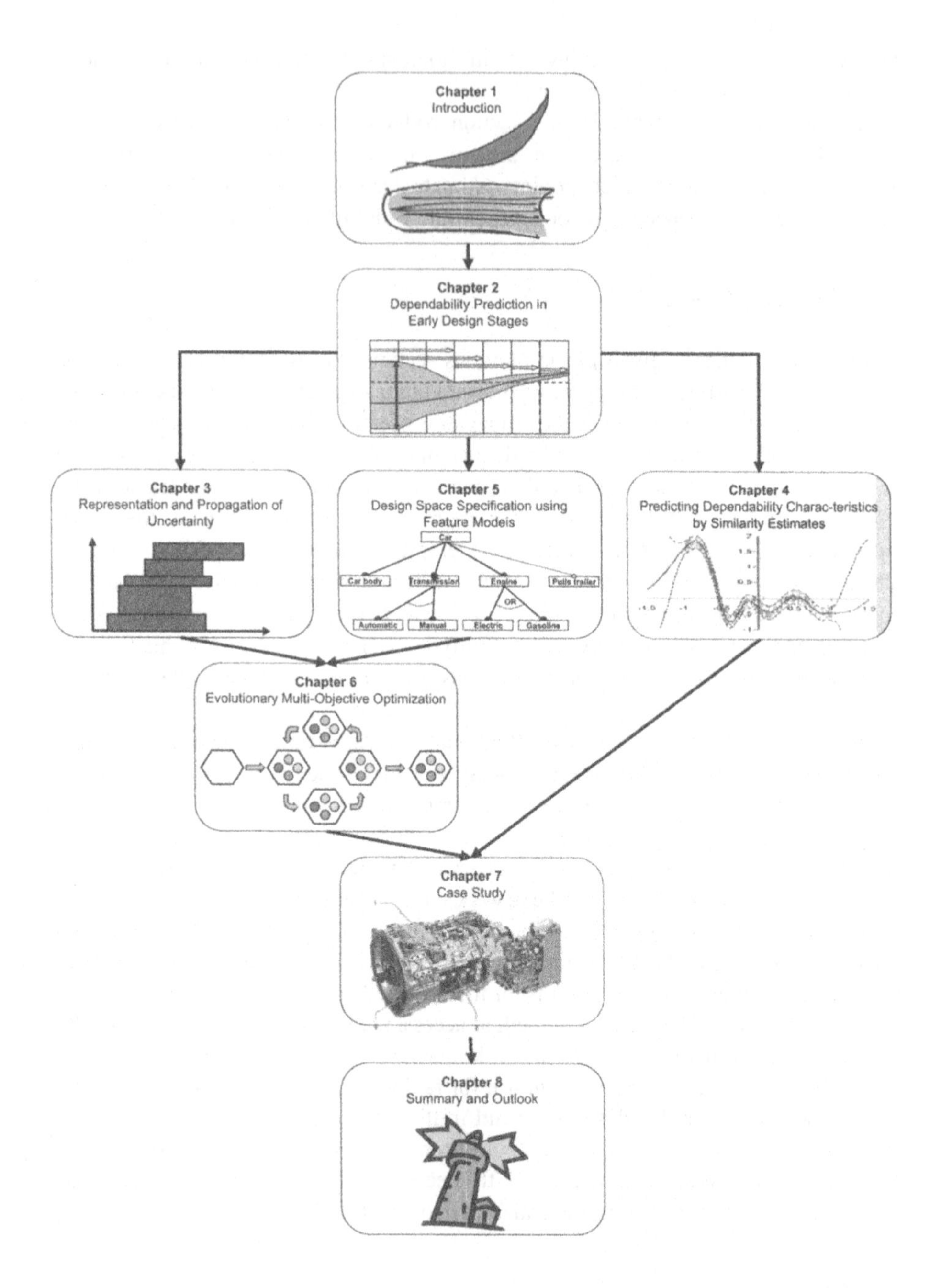

Fig. 1.1. Chapter overview

the necessary means of conducting an uncertainty study in the Dempster-Shafer representation. After introducing the ESReDA uncertainty framework which serves as a baseline for the application, the main concepts of the Dempster-Shafer theory of Evidence are introduced. It is shown, how expert estimates can be represented and merged into a multivariate belief function. The section also yields more advanced concepts such as measures of uncertainty (randomness, imprecision) and DST-based sensitivity analyses, which are still at the dawn of application.

In chapter 4, a new similarity-based expert elicitation procedure for failure distributions prediction is presented. Experts estimate relations to in-service components with known dependability characteristics. Neural networks and Gaussian processes, being two representative learning strategies from the field of computational intelligence, are used to capture relations such as "higher quality" and "lower wear out" to predict cumulative failure distributions (CDF) of new components including prediction uncertainty. The approach is validated with a case study from the automotive field.

In chapter 5, a new way of specifying design-for-dependability problems based on feature models is illustrated. Feature modeling, a modeling language describing sets of variants which originated from software engineering, can help the engineer to rapidly specify sets of design alternatives of future systems. In this thesis, feature models are utilized as a description language for sets of fault trees and sets of reliability block diagrams. An interface, which allows arbitrary optimization algorithms to screen this set for optimal designs is formulated. With this specification method, it is possible to represent arbitrary redundancy allocation problems (RAP) which are commonly analyzed in literature, but also more complex and especially more realistic design spaces.

Chapter 6 covers the development of a new, efficient multi-objective evolutionary algorithm (MOEA) for optimization under uncertainty in the Dempster-Shafer domain. After describing MOEAs and the scarce works covering EAs for uncertain objective functions, a dominance criterion and a density measure for imprecise solutions is proposed and built into a novel MOEA. In combination with the specification method in chapter 5, this MOEA can be used as an efficient design screening method for EDS.

The developed methodologies are applied in a IEC 61508 SIL compliance test in an EDS (chapter 7). The uncertainty is quantified, aggregated and propagated in the DST representation. It is shown, how with DST, a robust compliance analysis can be achieved. After testing the compliance of the original system, a feature model of possible design alternatives is set up and screened by the new MOEA.

A brief summary is given and further directions are outlined in chapter 8. Particular attention is paid on the evaluation of DST as a method for practical uncertainty studies. Its strengths and weaknesses are critically evaluated and suggestions on when to apply DST and when to revert to a probabilistic approach are given.

2 Dependability Prediction in Early Design Stages

2.1 The Mechatronic Project Cycle and Its Demand on Dependability Prediction

Mechatronic reliability, mechatronic safety or in its more general form mechatronic dependability [90] emerged from the need for comprehensive reliability and safety analyses of mechatronic products. The term "Mechatronics" was coined in 1969 by the Japanese engineer Tetsuro Mori for denoting an engineering field that unites several engineering domains such as mechanics, electronics and software/controls. In [34], mechatronics is defined as:

Definition 2.1 (Mechatronics). *Mechatronics is the synergistic combination of mechanical engineering, electronics, controls engineering, and computers, all integrated through the design process.[34]*

As can be seen in this definition, it is especially emphasized that all domains are *integrated* in the design process. Figure 2.1 illustrates the concept of mechatronics as a synergy of these four engineering domains. The outer ring shows some of the main application areas of mechatronic technology. The engineering discipline of mechatronics has become increasingly popular in the last decade with the ongoing integration of software into mechanic and electronic components and the resulting rapid growth of functionality for the customer. The proceeding integration of these domains and especially the growing variety of functionality has led to an enormously increasing system complexity, requiring the combination and enhancement of product development strategies.

Integrated, domain-spanning design methods are constantly developed [114, 69], but quantification of dependability prediction seems to be right now not able to keep up with this trend. Still traditionally divided into software, electronics and mechanics, it is difficult to meet up the demand for synergetic mechatronic dependability predictions with the current methods and tools [81]. As a consequence, many products suffer from severe dependability problems. In the automotive area, one of the main mechatronic industry branches in Germany, dependability problems are still continuously growing. Figure 2.2 shows an increasing number of forced callbacks of cars sold in Germany

P. Limbourg: Dependability Modelling under Uncertainty, SCI 148, pp. 7–19, 2008.
springerlink.com

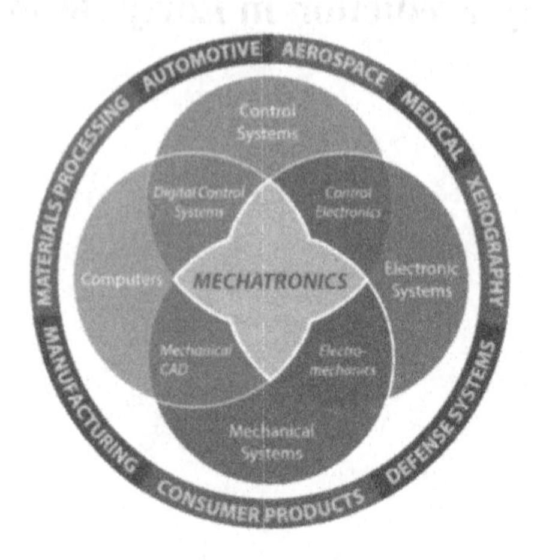

Fig. 2.1. Mechatronics, overview [138]

as issued by the German Kraftfahrt-Bundesamt (Federal Motor Vehicle Office) [92]. Moreover, these callbacks do not concentrate on one specific domain, but cover all domains touching mechatronics (with most callbacks being caused by mechanical problems). Therefore the need to integrate dependability prediction into the mechatronic design process is indisputable.

But perhaps the shifted requirements for the analyst caused by the changed development process are more crucial for dependability science than the increasing complexity and ongoing integration of technology. As an example, nuclear power plants may also be considered as mechatronic systems, and dependability in the nuclear sector has been at the center of attention from the very beginning of dependability prediction. Yet, these projects are huge endeavors with very small production quantities, where time constraints are less important and the dependability analysis budget is higher. In mechatronic development, projects cover whole product families and the time-to-market pressure is enormous. Mechatronic processes are highly structured and the applied process models require simulation and modeling tools that accompany the development process

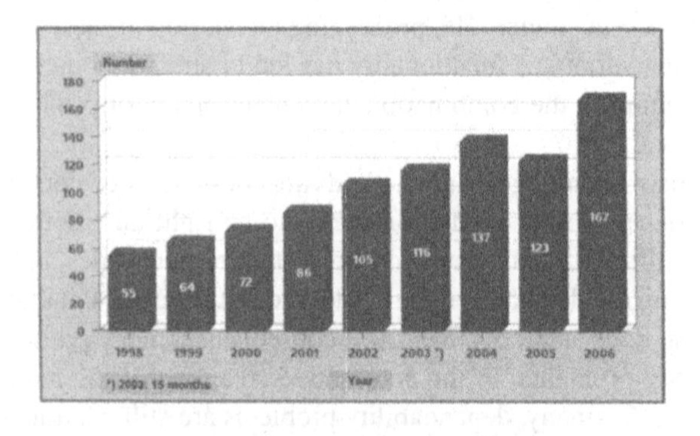

Fig. 2.2. Callbacks in German automotive industry, 1998-2006 [92]

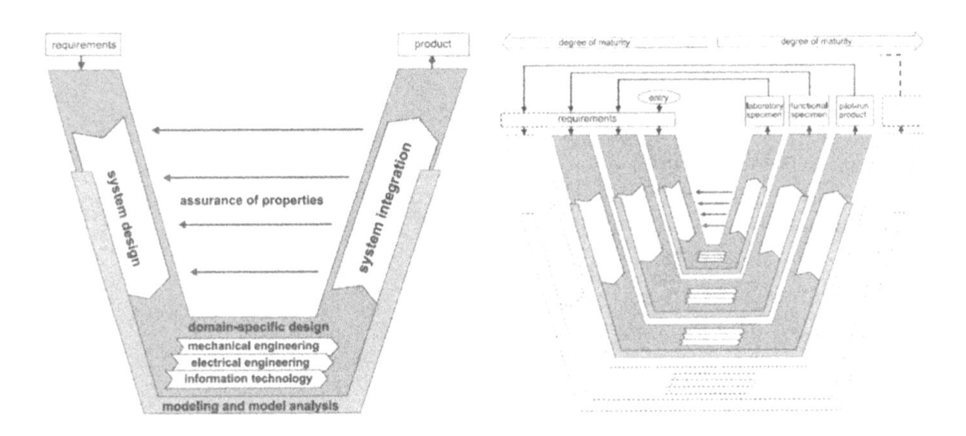

Fig. 2.3. Left: V-Model according to VDI 2206 [169]. Right: Iterative execution of the V-Model (macrocycle) with increasing product maturity [169].

from its very beginning [119]. Thus, an integration of dependability assessment into common process models such as the V-Model is a prerequisite to develop dependable products even under severe time constraints.

2.1.1 The V-Model: A Mechatronic Process Model

The V-Model is a process model that originated from software development. Introduced as a guideline by the German government, it has rapidly become one of the major process models in the IT sector. The V-Model XT [18] as the current version, is a very flexible process model or rather a "metamodel". In contrast to other process models, it defines activities and events, but does not restrict these events to be in a specific order. In contrary, the order may be fixed according to the project type. The V-Model XT defines the project types "incremental", "component-based" and "agile", thus permitting the tailoring of the process model to the project's specific requirements.

The need to integrate mechatronic methods into the V-Model stems from its practical relevance. It can be regarded as a quasi-standard for mechatronic projects in Germany, as it is proposed in the VDI guideline VDI 2206 "Design methodology for mechatronic systems" [169] as the process model to use. The incremental project type of the V-Model XT is the instance used in the VDI 2206, even if this project type is not explicitly defined in the guideline (figure 2.3, left). VDI 2206 structures the mechatronic development process into one or several iterative passes of the incremental V-Model as a macrocycle (figure 2.3, right), where each step contains one or more microcycles including the activities in the current design stage [62].

2.1.2 The Mechatronic Dependability Prediction Framework and the Integration of Dependability into the V-Model

In [13], the need for integrating dependability issues in the mechatronic V-Model is discussed. It is evident, that mechatronic projects can only benefit from a dependability

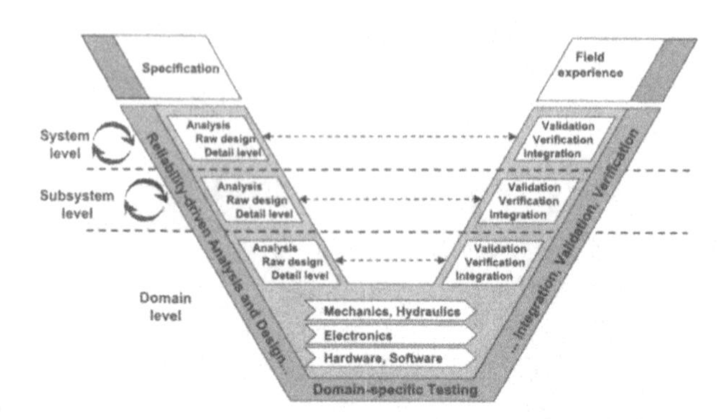

Fig. 2.4. Integration of dependability analysis into the V-Model [89]. Cycles indicate exemplary insertion points for microcycles.

prediction methodology in a full scale, which is completely integrated in the process. The V-Model as a standardized and widely-applied process model seems to be a good starting point for this endeavor. [89] proposes a framework to solve this problem. Because the V-Model requires a permanent validation and verification of the design process, dependability analysis is constantly performed throughout the project. Therefore, a "dependability layer" wrapping the whole development cycle needs to be integrated into the V-Model as shown in figure 2.4.

[91] provides the definition of mechatronic dependability together with an evaluation framework, which integrates predefinitions, evaluation and decision-making in a cyclic model. The definition combines the key factors of dependability for the mechatronic domain:

Definition 2.2 (Mechatronic Dependability). *The qualitative and quantitative assessment of degree of performance of reliability and safety related predefinitions taking into consideration all relevant influencing factors (attributes) [89].*

Figure 2.5 shows the accompanying framework. Activities are framed, effect directions are shown using arrows and the interfaces where results are provided are marked with dashed lines. The framework can be integrated as a microcycle into the V-Model as indicated by the loops in figure 2.4.

The light gray boxes represent the steps necessary for the integration of "qualitative and quantitative assessment" within definition 2.1. The assessment process depends on the *predefinitions* which determine target, purpose and constraints of the analysis. Analysis outputs are only meaningful with respect to the given predefinitions. Predefinitions may be separated into five different types:

Unit. Entity with fixed boundaries (system / component) which is part of the evaluation.
Time period. Time period for which the dependability is measured.
Requirements. Dependability requirements of the unit.
Criteria. Description forms, characteristics, measures of dependability.
Acceptance. Optional acceptance thresholds for a criterion.

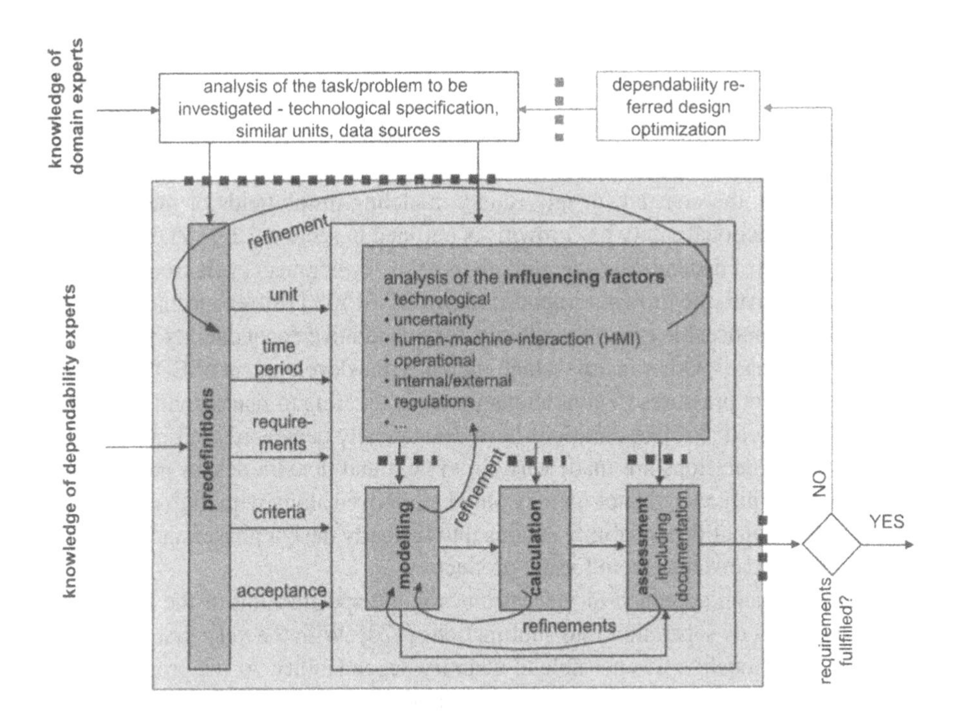

Fig. 2.5. Mechatronic dependability evaluation framework [91]

Predefinitions are derived by a collaborative effort of a *dependability expert* and one or several *domain experts*. Both roles are necessary, because in mechatronics, where several domains unite, the dependability expert with competence in dependability assessment may not have the overview over the domains. One or several domain experts with knowledge on the units of interest are needed to collaborate and generate a realistic dependability evaluation.

The evaluation itself consists of three steps. The first step is the *modeling* step, where models of the unit such as fault trees or reliability block diagrams are created. Based on the models derived, in the *calculation* step results are obtained, which are postprocessed, documented and refined in the *assessment* step.

The whole assessment process needs to take into account the relevant *influencing factors*, such as regulations by law, customers or technical communities, common-mode effects and interactions with humans (HMI) or other (eventually dependability-critical) systems. Beyond the scope of the evaluation process (light gray box), at the result interface, the results are checked against the requirements. Optionally, they can be used for dependability-driven design changes.

The framework shows the key aspects of dependability evaluation. Especially emphasized is the role of the domain experts whose integration into the evaluation process is crucial in mechatronics. Due to the cyclic shape of the model, which allows several iterations, it is fit for the integration into the V-Model as a microcycle. In [105], this is demonstrated based on the dependability analysis of ALDURO, an anthropomorphically legged robot [72].

2.2 Dependability in an Early Design Stage

In the recent past, the demand for uncertainty modeling in the fields of mechatronic reliability and functional safety has grown. As outlined in section 2.1.1, VDI 2206 demands integration of dependability prediction into the development cycle from the very beginning. Moreover, the growing importance of IEC 61508 [78] for mechatronic design has led to a noticeable growth in interest for quantitative dependability prediction over all project phases. Other factors which influence the development of EDS dependability methods are pressures to continuously reduce the time to market and to extend the range of products. Project durations have dramatically decreased, and most of the important design decisions are made early in system and domain design phases. Dependability flaws discovered later on may only be removed along with high costs. This favors the inclusion of dependability studies into the early design phases and the reuse of dependability knowledge from former products.

A common design approach in mechatronics is the specification of the basic system functions and its separation into subfunctions [128]. With the very first definition of the basic functionality, it is possible to quantify dependability. At the product interface, there are functions whose failure would be noticed by the customer [168]. System dependability models such as fault trees map physical events, subfunctions and part models to these functions (top events).

However, most companies rely solely on FMEA (Failure mode and effects analysis) in EDS, mainly without treatment of uncertainties and without systematic use of previous analyses. Meanwhile, with the increasing modeling power, the research focus moved from qualitative models without uncertainty treatment to quantitative prediction models analyzing and controlling the uncertainties, too. However, such methods can only penetrate the industrial practice if they provide a benefit, e. g. if quantitative results can be obtained rapidly with little additional modeling effort. Concurring with qualitative methods such as FMEAs, accuracy may be comparably low as long as this uncertainty is reflected in the outcome. In [82] and [104], the benefits of quantitative analysis as an aid for justifiable decision making between two or more design alternatives are shown.

A quantitative EDS dependability prediction method needs to provide a first dependability estimate as soon as possible and with low additional effort compared to qualitative models. Results may be approximate but need to conserve the parameter uncertainty. Following this demand, simple and efficiently usable prediction methods in an early project phase have to be developed. As dependability is predicted early, there is the risk of prediction error. The exact dependability is not predictable over the whole project cycle, and even good approximations remain unknown until later project stages. Figure 2.6 illustrates this behavior, plotting the dependability prediction including uncertainty depending on the project phase.

At the very beginning of the project, the only dependability data available stems from expert estimates and experiences from past projects. The system is still in its

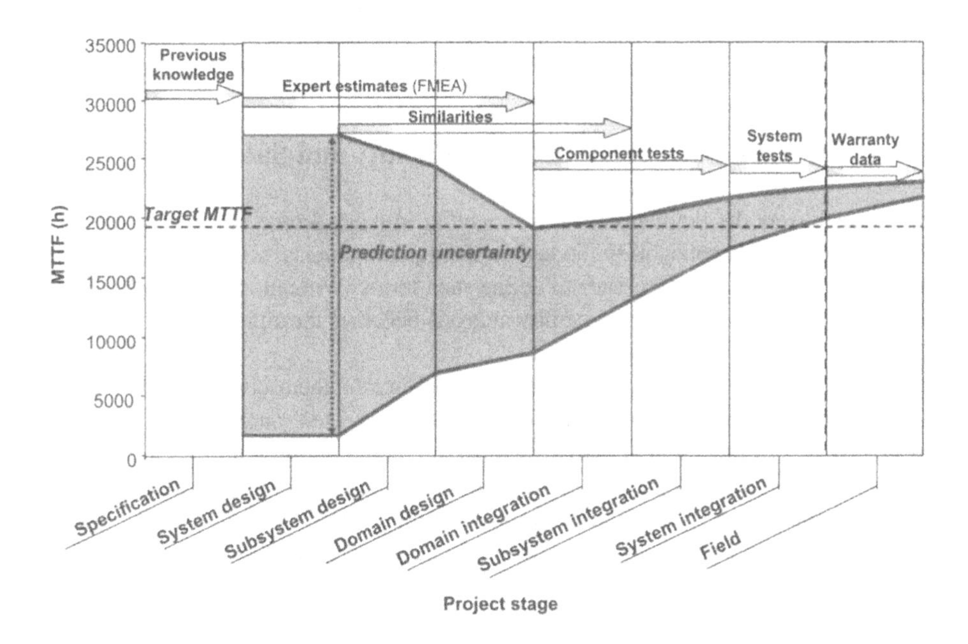

Fig. 2.6. Dependability prediction development in a project [104]

design phase and therefore the predicted dependability properties (in the illustration represented by the MTTF) are highly uncertain. Only wide bounds can be given on the estimated property. With the ongoing project, other data sources such as component similarity information, "handbook predictions" (e. g. the MIL-HDBK [163]) and finally test and field data reduce the prediction uncertainty.

However, dependability estimates are never certain until the last systems have been taken out of service and 100% of the field data is available. Thus dependability predictions need to reflect this uncertainty. Precise predictions therefore would disregard the imprecision of the data available and feign accuracy. Therefore, it is necessary to deal with uncertainties using uncertainty-preserving modeling methods. In [12], it is shown that uncertainty modeling in a reliability project does not only cost but may save resources by reducing the risk of taking the wrong decisions. This "value of information", commonly calculated in financial risk prediction is often neglected in reliability projects.

While dependability standards and guidelines applied in Germany neglect the necessity of uncertainty assessment, this is not the case for all European countries. The British military standard DEF Stan 00-42 [120] describes uncertainty in a R&M (reliability and maintainability) case. What makes DEF Stan 00-42 different from others is the explicit consideration of the dynamics in reliability knowledge during a project. The amount of information on the product's reliability is referred to as the "body of evidence". The body of evidence is an entity which includes all necessary information and can be considered as both information about the system and the environment. This body of evidence may change over time. Prediction accuracy will grow with product maturity, but system redesigns and introduction of new components / technologies may shrink the body of evidence again, thereby increasing the prediction uncertainty.

2.3 Definitions on Dependability, Reliability and Safety

This thesis covers the prediction of dependability in mechatronic systems. However, before setting up a dependability model containing components, system structure and failure probabilities, it is necessary to define such terms. Without an exact definition, the explanatory power of a dependability analysis becomes meaningless as the model may not be linked to a real-world system.

It is important that the given definitions do not contradict themselves. This motivates the use of the whole terminology from one, generally accepted source instead of creating a patchwork of incongruent ones. The most influential set of definitions in the area of dependability grounds on the work of Laprie [98]. Its successors [99] and [4] offer perhaps the most coherent terminology existing in the field. In the course of this thesis, the terminology of [4] will be used exclusively. The following three subsections offer a detailed description of the elements (section 2.3.1), attributes (section 2.3.2) and means (section 2.3.3) to attain dependability. All given definitions originate from this source.

2.3.1 Basic Definitions of Elements in Dependability Modeling

In system dependability, the object of interest is (obviously) a *system*. This system is located in an *environment* with which it can interact. This context can be defined as:

Definition 2.3 (System, environment). *A system is an entity that interacts with other entities, i. e. other systems including hardware, software, humans, and the physical world. These other systems are the environment [4].*

This definition, describing systems as entities interacting with the world is similar to the definition of a unit (Entity with fixed boundaries (system / component) which is part of the evaluation) in section 2.1.2. In this thesis, the term system will be utilized because of the more common use. If dependability is modeled, the system has a purpose, a function which should be fulfilled.

Definition 2.4 (System function). *The system function is what the system is intended to do [4].*

Definition 2.5 (System behavior). *The behavior is what the system does to implement its function and is described by a sequence of states [4].*

The specification of system functions and their separation into sub-functions plays a key role in the EDS of a mechatronic system [169]. Therefore, being defined very early, they can be integrated in dependability models from the start. The system service is the main element for further definitions of reliability and safety. It is important to notice that the system behavior may differ from the implementation of the system function. This is the scenario that needs to be avoided.

Definition 2.6 (Service, User, Service interface). *The service delivered by a system is its behavior perceived by its user(s). A user is another system that receives service from the providing system. The service delivery takes place at the service interface. The delivered service is a sequence of the provider's external states [4].*

Definition 2.7 (External system state). *The external system state is the part of the providing system's total state that is perceivable at the service interface. The remaining part is the internal state [4].*

System dependability prediction lives on the separability of a complex system into smaller building blocks. The discussion on the correct definition of the terms "system structure" and "component" is a more or less philosophical question, which is excluded from this thesis. Reference is made to [84] where the topic is covered in detail.

Definition 2.8 (System structure). *The structure of a system is what enables it to generate the behavior [4].*

Definition 2.9 (Component). *From a structural viewpoint, a system is composed of a set of components bound together in order to interact. Each component is another system or atomic [4].*

Definition 2.10 (Atomic component). *An atomic component is a component whose internal structure cannot be discerned, or is not of interest and can be ignored [4].*

As the recursive component-system definition shows, it is necessary to stop refining the model, once an agreeable level of detail is reached. This does not mean that atomic components are really atomic. It is just not necessary or worth the expense to particularize the model further. For the definition of the aspects of dependability in section 2.3.2, it is necessary to define correct service and outage:

Definition 2.11 (Correct service). *Correct service is delivered when the system implements the system function [4].*

Definition 2.12 ((Service) Outage). *Service outage is the period of delivery of incorrect service [4].*

This view is important for the later introduced Boolean models (section 2.4). It supports the abstraction from a complex system to a Boolean model. Even if the system itself is not Boolean, there is the possibility to define a threshold or boundary condition which separates correct service from incorrect service, hence introducing a Boolean property. Failures are defined as events, transitions from correct service to incorrect service. Opposite to this, service restorations are transitions from incorrect to correct service.

Definition 2.13 ((Service) Failure). *A (service) failure is an event which occurs when the delivered service deviates from its correct service [4].*

Correct service is delivered if the system with its behavior implements the system function. Therefore service failure implies that one or several external states deviate from the correct state. Such specific deviations from the correct service, e. g. overly large time delays are referred to as errors. Faults are causes for errors and thus for failures:

Definition 2.14 (Error). *An error is the deviation of an external state from the correct service state [4].*

Definition 2.15 (Fault). *A fault is the adjudged or hypothesized cause of an error [4].*

2.3.2 Dependability and Its Attributes

In the last subsection, all mandatory elements for a thorough definition of dependability, reliability and safety were given. The central property of a system according to [4] is dependability. They illuminate the relationship between dependability and dependence as the dependability of a system should reflect the dependence placed in it.

Definition 2.16 (Dependability). *The ability [of a system] to avoid service failures that are more frequent and more severe than is acceptable [4].*

Definition 2.17 (Dependence, Trust). *Dependence of system A on system B: The extend to which A's dependability is affected by that of system B [4].*

A system on which nothing in its environment depends does not need to be highly dependable. The same system in another environment may need to be highly dependable. An accelerometer does not need to be very dependable for image stabilization in a camcorder. However, it needs to be highly dependable if being used in an airbag sensor system. Dependability is interpreted as an integrating concept subsuming five attributes:

Availability. Readiness for correct service
Reliability. Continuity of correct service
Safety. Absence of catastrophic consequences on the user(s) and the environment
Integrity. Absence of improper system alterations
Maintainability. Ability to undergo modifications and repairs

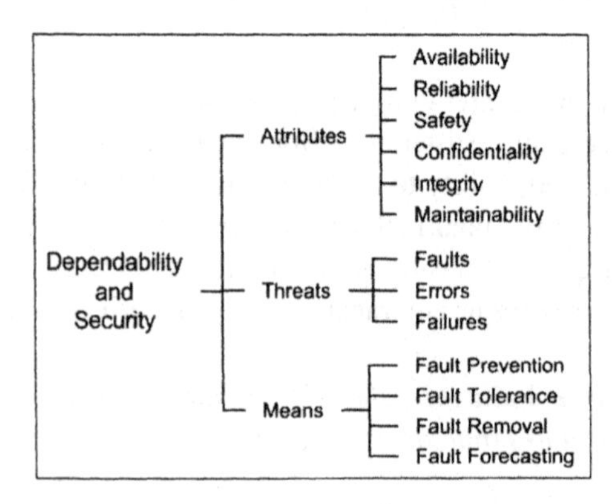

Fig. 2.7. The dependability and security tree [4]

Figure 2.7 illustrates the connections between dependability and its attributes, threats and means to achieve it. Attributes and threats have already been defined, means to achieve dependability will be introduced in section 2.3.3. The viewpoint represented by figure 2.7 contradicts earlier works such as [98] where the main constituents were reliability, safety, security and fault tolerance. Security is no longer interpreted as an attribute of dependability, but as a similar concept sharing most of its attributes. Security is the composition of availability, integrity and confidence (Absence of unauthorized system alterations). Fault tolerance is not a part of but a means to achieve dependability.

2.3.3 Means to Attain Dependability

To obtain a high dependability, a great number of different means have been developed, which according to [4] can be divided into four groups:

Fault prevention. Means to prevent the occurrence or introduction of faults (e. g. oversizing, design patterns).

Fault tolerance. Means to avoid service failures in the presence of faults (e. g. redundancy).

Fault removal. Means to reduce the number and severity of faults (e. g. functional testing).

Fault forecasting. Means to estimate the present number, the future incidence, and the likely consequences of faults (e. g. fault tree analysis).

Fault prevention and fault tolerance play key roles in mechatronic systems. All domains have developed mature concepts for fault prevention that are either operational (oversizing) or developmental (software design patterns) and fault tolerance. Fault removal is very important in later project stages and there are sophisticated methods for testing and fixing. However, fault removal is costly and may cause system redesign, motivating the fourth category, fault forecasting.

Fault forecasting emerges from the desire to "fix faults before they emerge". Dependability modeling, prediction and assessment are often traded as synonyms for fault forecasting. Especially in EDS, where faults cannot be observed, fault forecasting is important. It leads to the prediction and thus to the prevention of the introduction of faults before the first prototype is produced. Fault forecasting grows in importance with the size of the project and its financial constraints. Fault prevention is only economically justifiable up to a certain amount [80]. Fault forecasting thus can be seen as the mean to discover possible dependability flaws in the design and to find the most efficient locations for fault prevention and fault tolerance. The main part of this thesis falls into the category of fault forecasting, namely quantitative dependability analysis based on DST.

2.4 Boolean System Models

This section introduces shortly Boolean system models [146, 171], the class of dependability models used throughout this thesis and their linkage to the definitions of [4]. Common types of system functions for Boolean models are reliability block diagrams

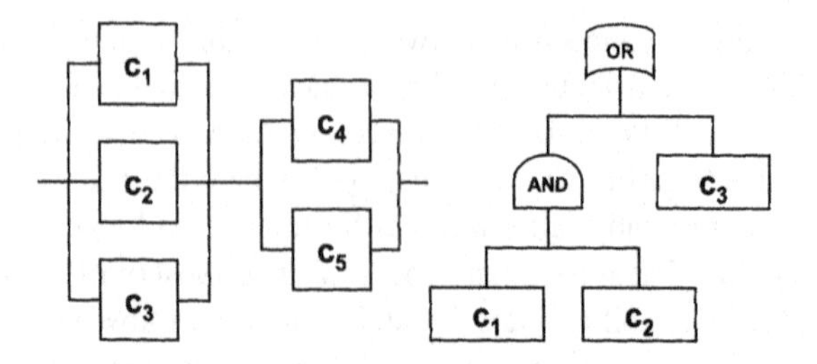

Fig. 2.8. Boolean system models. Left: Reliability block diagram. Right: Fault tree.

[88], fault trees and event trees (see figure 2.8). Boolean systems, especially fault trees are a very popular modeling approach for reliability and safety in practice [97, 45].

The objective of system dependability analysis is the calculation of system dependability properties (probability of critical failure, lifetime) from component dependability properties. The system state on which the system behavior (sequence of states, def. 2.5) and thus the service (the behavior perceived by the user, def. 2.6) depends, is expressed by a variable s_s. The basic premise of system dependability prediction is that s_s can be obtained by knowing the current states of the components forming the component state vector c_s and the system structure, described by the system structure function ϕ_s. The following notation will be used throughout this thesis:

Definition 2.18 (System function). *Let C_s be the set of all possible component state vectors and S_s the set of all system states. A system function $\phi_s : C_s \to S_s$ is a function $\phi_s(c_s) = s_s$ which maps each component state vector $c_s := (c_1, ..., c_n)^T$ to a system state s_s.*

Depending on the domains of S_s and C_s, system dependability models may be classified into different categories. The most common type of models are Boolean models, which restrict the system and component state spaces to $\phi_s : C_s \subseteq \mathbb{B}^n \to S_s \subseteq \{0,1\}$. The system states $s_s = 0$ and $s_s = 1$ are commonly linked to "service outage" and "correct service" and thus the model is consistent to the taxonomy of [4]. If $\phi_s : C_s \subseteq \mathbb{N}^n \to S_s \subseteq \mathbb{N}$, then multi-state systems emerge [129]. For $\phi_s : C_s \subseteq \mathbb{R}^n \to S_s \subseteq \mathbb{R}$, continuous systems [60] are given. Both multi-state and continuous systems are less commonly used than Boolean systems.

Boolean models mainly serve as an underlying model for a probabilistic description of the system. Atop of the Boolean model, there is a function ϕ_{PS}, mapping the component failure probabilities to the system failure probability. Therefore most dependability engineers consider the probabilistic level 1 uncertainty representation the "true" model. The probability of a component $i = 1, ..., n$ to fail is described by a failure probability $p_i \in [0,1] := P(c_i = 0)$. The component probability vector p represents the combination of all component failure probabilities:

$$p := \begin{pmatrix} p_1 \\ \vdots \\ p_n \end{pmatrix} \tag{2.1}$$

The system failure probability p_s can then be obtained via probabilistic computation from p. This function ϕ_{PS} is denoted as:

$$\phi_{PS} : [0,1]^n \rightarrow [0,1] \tag{2.2}$$

$$\phi_{PS}(p) = p_s \tag{2.3}$$

Thus, the vector p represents the input parameter of the system dependability model in a probabilistic representation. In real life, p is never accurately known and in fact there are uncertainties on a large scale that disturb the exact knowledge on p. This thesis will therefore illustrate how to form an uncertainty model around p and project this uncertainty onto the output.

3 Representation and Propagation of Uncertainty Using the Dempster-Shafer Theory of Evidence

This chapter collects fragments from current research to form a coherent framework for uncertainty representation and propagation in the Dempster-Shafer theory of evidence (DST). While there are some aspects of DST that have been brought into practice, many other methods important in probabilistic modeling, such as uncertainty measures and sensitivity analyses were only proposed but not transferred to real-world problems. Especially in the field of dependability prediction, the application of DST is still in its fledgling stages. Hence, the purpose of this section is not only to describe the state of the art but to present and introduce new tools necessary for a DST-based uncertainty analysis in a dependability scenario. All methods described are implemented in an open source toolbox (Imprecise Probability Toolbox for MATLAB, [103]).

The chapter first covers the various types and sources of uncertainty (section 3.1). Section 3.2 describes the ESReDA framework on uncertainty modeling, a guideline for uncertainty modeling in practice, followed by an introduction to the Dempster-Shafer theory of evidence and a literature review of its applications in dependability (section 3.3). The following sections cover in more detail the aggregation of several belief functions describing one component (section 3.4), the generation of a joint belief function using copulas (section 3.5) and the propagation of this joint belief function through a system model (section 3.6). The chapter ends with sections covering uncertainty measures (section 3.7) and sensitivity analysis (section 3.8) in the framework of Dempster-Shafer theory. Throughout this chapter, a simple "illustrative example" will help to exemplify the modeling process, which is the base of the case study in chapter 7. This chapter concludes with some recommendations on when to use DST and when to revert to a probabilistic uncertainty model (section 3.9).

3.1 Types and Sources of Uncertainty

In Boolean dependability models, the system failure probability p_s is obtained from the component failure probabilities x and the system function ϕ (see section 2.4). In common dependability modeling practice, the vector p is represented by deterministic values. It is assumed that the analyst has perfect information on p.

However, EDS system dependability prediction, as the name implies, is a prediction methodology. One can not be absolutely certain about the component failure probability

P. Limbourg: Dependability Modelling under Uncertainty, SCI 148, pp. 21–51, 2008.
springerlink.com

vector p, its variability or even the system function. Lack of information, conflicting information or the limited ability of the model to reflect the real world prevent this. Therefore uncertainty-preserving models can be used to represent uncertainty in p. Because of the limited degrees of freedom to represent uncertainties, it is difficult to express sources other than physical variability even in a probabilistic framework. Other representations of uncertainty such as imprecise probabilistic models are alternatives. However, before turning the scope to the mathematical representation of uncertainty, terms such as information, evidence and certainty have to be defined and possible sources of these uncertainties have to be identified.

Two of the most controversial discussions in uncertainty analysis are on the distinction of uncertainty into several types and on the classification of possible sources of uncertainty. A popular distinction [74, 127] is the separation of uncertainty into two categories: aleatory uncertainty and epistemic uncertainty. Uncertainty in Dempster-Shafer theory can be represented in various ways and can therefore support this classification. Aleatory uncertainty arises from variability of the system or environment considered. Oberkampf [125] defines aleatory uncertainty as:

Definition 3.1 (Aleatory uncertainty). *Aleatory uncertainty is the inherent variation associated with the physical system or the environment under consideration [125].*

Aleatory uncertainty occurs in random experiments such as dice throws or chaotic system behavior. It is often referred to as irreducible uncertainty and characterized by a random quantity with known distribution. The exact value will change but is expected to follow the distribution. A simple example for aleatory uncertainty is the uncertainty about the outcome of a coin toss $x \in \{0,1\}$. The observer is uncertain about head $x = 0$ or tail $x = 1$ of a single throw. Therefore it is usual to describe a coin toss with a random variable $\widehat{X}$. However, he is quite confident that each of the numbers will occur with a probability $P(\widehat{X} = 0) = P(\widehat{X} = 1) = 1/2$. Infinite repetitive executions of this experiment will lead to the observance that the relative frequency converges to the given probability values.

On the contrary, epistemic uncertainty describes not the uncertainty about the outcome of some random event due to system variance but the uncertainty of the outcome due to lack of knowledge or information. Thus, [125] define epistemic uncertainty as:

Definition 3.2 (Epistemic uncertainty). *Epistemic uncertainty is uncertainty of the outcome due to any lack of knowledge or information in any phase or activity of the modeling process [125].*

Definition 3.2 shows the important distinction between these two types of uncertainty. Epistemic uncertainty is *not* an inherent property of the system. A gain of information about the system or environmental factors can lead to a reduction of epistemic uncertainty. Therefore a common synonym for epistemic uncertainty is reducible uncertainty. Imprecision is also often used in a similar context. In [155], imprecision is defined as:

Definition 3.3 (Imprecision). *Imprecision is something that is not clear, accurate, or precise, not accurately expressed or not scrupulous in being inexact [155].*

Focusing again on the dice example, in a second scenario the coin could be biased and information about the exact probabilities is therefore imperfect. The observer might expect that the probability is limited by $P(\widehat{X}=0) \in [0.25, 0.75]$ or may itself be described by a second-order probability on $P(\widehat{X}=0)$. Of course, the coin follows a distribution and if multiple throws were carried out, evidence would grow. After an infinite number of experiments, the probability observed could be $P(\widehat{X}=0)=0.4$. However, before doing so, there is no evidence for assuming any possible distribution without neglecting that reality may be anywhere else. Hence, epistemic uncertainty is the inability to model reality.

Epistemic uncertainties occur in small or large amounts in almost all system dependability models, especially in EDS. It is either impossible to carry out enough experiments for gathering informations about the quantity of interest, or it is possible, but the required resources (time, budget, manpower...) are excessive and pose constraints to the project. Traditional dependability engineering applications tend to model only aleatory uncertainties, which can lead to under-/overestimations of the system dependability (section 2.2).

While the application of probability theory is the reasonable choice for modeling aleatory uncertainties, this is not necessarily the case for epistemic quantities. The behavioral or Bayesian interpretation associates probabilities with degrees of belief and thus tries to transfer probability theory to belief modeling [153]. This opposes to other theories such as fuzzy sets [182] and the Dempster-Shafer theory of evidence. Dempster-Shafer theory will be the representation used in this thesis and is introduced in section 3.3.

After the categorization of the different types of uncertainty, it is valuable to look on the possible sources that cause uncertainty (both aleatory and epistemic). In [170], fourteen types of uncertainty sources are given:

Lack of information. There is not enough evidence concerning the quantity of interest.

Conflicting information. Different sources of information are incongruent (e. g. Weibull shape parameters $\beta \in [1.2, 1.5]$ and $\beta \in [2, 2.3]$ estimated from two tests).

Conflicting beliefs. In contrast to the conflicting information, conflicting beliefs are not based on evidence but on expert opinion. The experts may have access to different sources of information or assess the information in different ways.

Information of limited relevance. The information that the model is based upon is only scarcely usable for prediction.

Physical indeterminacy. The quantity of interest is not precisely predictable, even with infinite information.

Lack of introspection. Experts do not have time or resources to reduce the amount of uncertainty in their estimate. Walley [170] calls this the "cost of thinking".

Lack of assessment strategies. Relevant information is available, but it is not straightforward to introduce it in the model (e. g. textual reports). If the data is transformed, uncertainty rises.

Limits in computational ability. Imprecision can be introduced by lack of computational power such as a limited amount of Monte Carlo simulations.

Intractable models. The real model might be inconvenient or too complex. If replacing it by a simpler one, uncertainty rises.

Natural extension. Additional uncertainty can be introduced because of the ambiguity of uncertainty propagation through the system model.

Table 3.1. Sources of uncertainty and classification into indeterminacy and incompleteness[170]

Indeterminacy	Incompleteness
Lack of information	Lack of introspection
Conflicting information	Lack of assessment strategies
Conflicting beliefs	Limits in computational ability
Information of limited relevance	Intractable models
Physical indeterminacy	Natural extension
	Choice of elicitation structure
	Ambiguity
	Instability
	Caution in elicitation

Choice of elicitation structure. Depending on the experience of the assessor and the assessment technique, there may be a limit on the accuracy of the elicitation.

Ambiguity. The "fuzzy" characteristics of linguistic estimates such as "pretty likely" and "almost impossible" introduce uncertainty if translated to quantitative values.

Instability. Using different elicitation methods or even the same method twice may lead to different assessments, because beliefs can be unstable (e. g. underlying information is remembered or weighted different).

Caution in elicitation. Engineers may give higher imprecision ranges than necessary if they fear to be blamed for wrong estimates. Imprecision may also be introduced, if data is voluntarily held back, or the imprecision is artificially increased. This may be the case if dependability data is exchanged between companies.

Walley [170] classifies these sources into uncertainty caused by *indeterminacy* and by *incompleteness* (table 3.1). Indeterminacy reflects limitations of the available information. Incompleteness is caused by a simplifying representation which permits the usage of only a partial amount of information available.

3.2 The ESReDA Framework on Uncertainty Modeling

As the number of applications covering uncertainty modeling in EDS dependability prediction is limited, it is a convenient strategy to adapt the knowledge from other domains with more experience in this field. The ESReDA Framework on Uncertainty modeling [50] is a valuable tool for integrating uncertainty analysis into mechatronic dependability prediction. Derived from a set of industrial case studies spanning several engineering domains, it tries to subsume commonalities and to work out key aspects of uncertainty studies in industrial practice. It provides a general scheme for such studies which is easy to follow.

The need for uncertainty modeling occurs in various different areas of industrial practice (e. g.[139], [5] and [140]). Everywhere a model describes some dependability properties, uncertainty is involved [124]. According to [50], quantitative uncertainty assessment in industrial practice involves (figure 3.1):

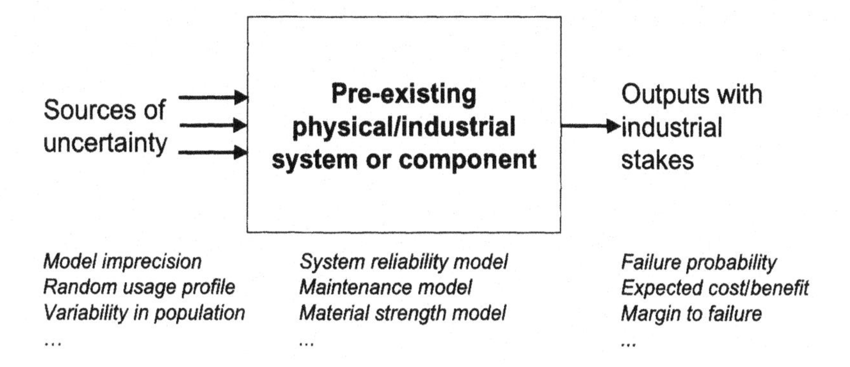

Fig. 3.1. Quantitative uncertainty assessment in industrial practice (adapted from[50])

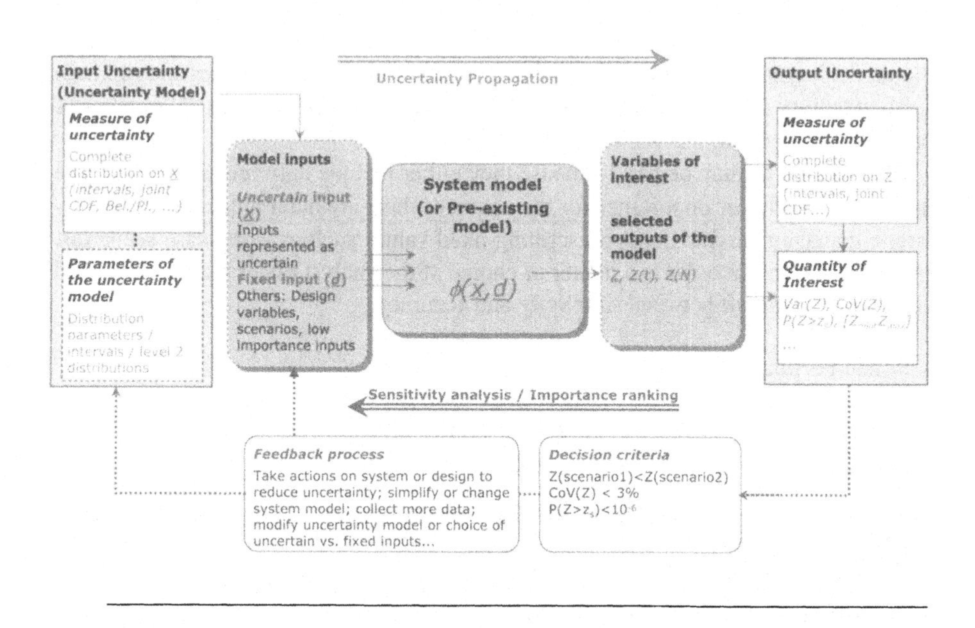

Fig. 3.2. The ESReDA framework on uncertainty modeling [50]

- A physical or industrial system or component represented by a (mainly pre-existing) system model,
- Sources of uncertainty which affect this model,
- Industrial stakes motivating the uncertainty assessment.

Various sources of uncertainty exist, which have been presented in section 3.1. Depending on the domain vocabulary, they may be named as "aleatory/epistemic", "irreducible/reducible", "lack of data", "variability" or "measurement error". They influence the system model in various ways. Three influence points for uncertainties are distinguished: uncertainty on the model inputs, errors on the model output and model uncertainty, uncertainty on the adequacy of the model itself. If model uncertainty can be integrated in uncertainty studies is an open question discussed in [124].

Industrial stakes are e. g. customer requirements, financial and economic optimizsation and regulatory certification. They are the purpose of the system model. The output of the model supports and helps handling those stakes in a quantitative way within the decision-making process. This system model forms the core of the ESReDA framework (figure 3.2), wrapping around the parts necessary to process the uncertainty.

System model

A system model such as the one given in figure 3.1 represents the core of an uncertainty study. Formally the model links the output variables of interest (noted z) such as the system state and a number of uncertain input variables $x := (x_1, ..., x_n)^T$ and fixed parameters d through a deterministic function $z = \phi(x, d)$. Some variables x are uncertain and will be subject of the uncertainty model while others, denoted as d are fixed. ϕ can be a simple Boolean function such as a reliability block diagram or a complex computational fluid dynamics simulation and thus may show arbitrary properties and runtime. In dependability models, the dimension of x can be large depending on the number of components described.

Other inputs (d) may be fixed because they either include only few uncertainty, the uncertainty is of lesser importance for the output or there are outer influences (e. g. regulatory, management decisions) prescribing fixed values such as worst-case scenarios. The separation into x and d is a matter of choice of the analyst rather than a theoretical discussion, a trade-off between simplicity and accuracy.

Goals and Feedback Process

[50] states that uncertainty studies in practice follow one or several goals of the following types:

1. Demonstration or guarantee of compliance with a threshold (e. g. SIL compliance [78]).
2. Comparison of relative performance / optimization (e. g. design comparison).
3. Further understanding of the uncertainties' influences (e. g. for test planning).
4. Calibration or simplification of a model such as reducing / fixing the number of inputs.

Several goals can possibly be followed simultaneously or sequentially during one study. A dependability model in an EDS may first follow the purpose of goal 3 (understanding the critical uncertainties) and goal 2 (finding a reliable and safe design) to be later on transferred to goal 1 (compliance with safety regulations / customer reliability specification).

The results of the analysis may trigger one or several feedback processes, such as the adjustment of the design to comply with a threshold (goal 1) or the selection of the best design alternative (goal 2).

Quantities of Interest

The treatment of the model output may change along with the goals of the study. However, they will implicitly or explicitly involve "quantities of interest" on the output variables. Examples for quantities of interest are:

- Probability of exceeding a threshold.
- Expected values of the variables of interest.
- Confidence intervals on the variables of interest.

In case of reliability and safety analysis, common quantities of interest include the probability of failure or the MTTF. If testing the compliance to a SIL, the exceedence probability of the SIL threshold could be a quantity of interest.

There may be various types of requirements on the system involving quantity of interest such as:

- Less than 3% uncertainty on an output variable of interest.
- Failure probability less than 1E-7 with 95% confidence.
- A better performance of alternative A compared to alternative B with 95% confidence.

Some of these requirements may be explicitly defined, some rather fuzzy, depending on the application area. In any case, one or several quantities of interest on the output variable are the results that will be used in the further decision-making process.

Quantities of interest are not directly obtained from the output variables, they are derived from measures of uncertainty. In a probabilistic framework, PDFs or CDFs can be considered as measures of uncertainty, in the DST framework, a mass distribution plays this role.

Uncertainty modeling and Propagation

The main part of the framework comprises of the uncertainty modeling and propagation phases. Depending on the measures of uncertainty and the quantities of interest chosen, uncertainties can be modeled in different ways. In a probabilistic representation, they may be represented by a joint PDF on x. The joint PDF is commonly simplified either by using independence assumptions or dependency models to break down the multivariate PDF to univariate distributions, which are often represented by parametric functions such as Gaussian or Weibull distributions. In the DST case, the multivariate PDF is replaced by a joint BPA with similar simplification possibilities (decomposition into marginal distributions). The uncertainty representation is derived from direct observations such as field data, expert estimates, physical arguments or indirect observations such as similarity information. The process of gathering and building the uncertainty model may be costly. However it is the most important step in the chain and requires a lot of experience and care.

The uncertainty model is propagated through the system model to create measures of uncertainty on the variables of interest (e. g. a PDF on $z = \phi(x,d)$ from a PDF on x and a vector d). Depending on the complexity of ϕ, the time constraints and the quantities of interest, a large set of methods such as different Monte Carlo sampling strategies [117] can be used. In the sensitivity analysis step, importance measures of the inputs x on a quantity of interest on z are derived.

3.3 The Dempster-Shafer Theory of Evidence

The Dempster-Shafer theory of evidence (DST) was first described by Dempster [42] and extended by Shafer [150]. The term "evidence theory" is often used as a synonym. DST is a mathematical representation that belongs to the class of imprecise probabilities [85]. All imprecise probability theories include the probabilistic representation as a special case, but add further means to represent uncertainty. This renders the application of DST to EDS dependability prediction especially tempting, because it is necessary to model several sources of uncertainty (lack of knowledge, conflicting beliefs, etc). It is not said that for all sources, the probabilistic representation is the most adequate [157]. A more powerful representation therefore could be beneficial for a thorough uncertainty quantification. Detailed overviews on DST and imprecise probabilities can be found in [170], [43] and [86]. The Society of Imprecise Probabilities and Their Applications (SIPTA) hosts a repository on imprecise probabilities [179] that includes tutorials and recent developments in this field.

3.3.1 Dempster-Shafer Theory in Dependability Modeling

DST was originally applied in artificial intelligence and sensor fusion with focus on the aggregation of evidence, e. g. for expert systems [151]. In the field of dependability prediction, imprecise probabilities are not yet in common use. The limited work available can be grouped into two categories. On the one hand the theoretically-oriented work of Utkin, Coolen and others [31] and [166]. While these works may be theoretical foundations, applications to larger real-world problems has not been given yet. Most of the approaches focus on the formal representation of system dependability prediction and decision-making under minimal knowledge in the DST representation. They involve high modeling and computational complexity, e. g. [164], [165] and [167]. Other dependability applications in this category involve [14], where the lifetime of a two-component system is obtained under minimal knowledge on the dependency pattern.

The second category, on the other hand, is a range of introductions to and applications of the basic concepts of DST. Most studies focus on the use of uncertainty propagation in the DST representation illustrated in different case studies. Exemplary works are the introduction by [133] and applications by [65] and [23]. [21] border on the reliability of a fault tree system model with interval arithmetic (which can be interpreted as a special case of DST) for fault tree computation. [104] use DST in a reliability prediction process under uncertainty. Other works such as [64], [143], [144] and [134] focus on reasoning and aggregating evidence in reliability, safety and maintainability scenarios. [24] covers imprecise reliability growth models. According to this literature overview, there is no case study or practical scenario in dependability or safety analysis,that uses the full modeling possibilities of DST such as dependency modeling, uncertainty measures and sensitivity analysis.

However it will be wrong to assume that everywhere in the area of uncertainty modeling, DST is only of academical importance. The epistemic uncertainty project, a large scale endeavor (e. g. [54]) of the Sandia National Laboratories in 2002 [178] aimed at exploring the use of DST in real uncertainty studies. The project culminated in a workshop of renown experts, whose outcome, a set of case studies (e. g. [162]) for solving benchmark problems for representing and propagating both epistemic and aleatory uncertainty based on DST [56, 71], motivated further research in this field. In several uncertainty studies, parts of the proposed methods were successfully realized. [93] propagates and aggregates evidence in large-scale climate models, [10] in physical models of ripening processes and [46] in risk assessment. Meanwhile sensitivity analysis was initially investigated in the DST scenario [58]. Thus, one of the main aims of this thesis is to transfer and extend this modern DST approach to the field of dependability prediction, so that the uncertainty representation benefits of DST can be used to gain a better image of the product dependability under uncertainty.

3.3.2 Foundations

In probabilistic modeling, uncertainty regarding a variable x is represented by a random variable $\widehat{X}$. This means that a mass m can be put on each value $x \in X$, expressing the probability $P(\widehat{X} = x)$. This mass function is given as:

$$m : X \rightarrow [0,1] \tag{3.1}$$

$$\sum_{x \in X} m(x) = 1 \tag{3.2}$$

In discrete probabilistic models, m defines a probability distribution and $P(\widehat{X} = x) = m(x)$. It is important to outline this well-known equality, because it will not hold in the case of DST.

From m, the probability of x to be in interval $\underline{\overline{x}} := [\underline{x}, \overline{x}]$, $P(\widehat{X} \in \underline{\overline{x}})$ can be obtained by adding over all mass values:

$$P(\widehat{X} \in \underline{\overline{x}}) = \sum_{x \in \underline{\overline{x}}} m(x) \tag{3.3}$$

As a special case of eq. 3.3, the cumulative distribution function (CDF) $F(x) := P(\widehat{X} \leq x)$ is completely determined by m. In the discrete probability representation, a mass is defined for each possible value of $x \in X$ (eq. 3.2). Dempster-Shafer mass functions on the real line are similar to discrete distributions with one important difference. The probability mass function is not a mapping $m : X \rightarrow [0,1]$ but instead a mapping $m : \mathscr{P}(X) \rightarrow [0,1]$, where masses are assigned to subsets of X instead of elements in X. In Dempster-Shafer theory, the analogon to a random variable $\widehat{X}$ is a belief variable $\widetilde{X}$, which can be described by its basic probability assignment (BPA) such as a random variable by its probability distribution.

Definition 3.4 (Basic probability assignment). *A basic probability assignment (BPA) m is a mapping* $m : \mathscr{P}(X) \rightarrow [0,1]$ *provided:*

$$m(\emptyset) = 0 \tag{3.4}$$

$$\sum_{A \subseteq X} m(A) = 1 \tag{3.5}$$

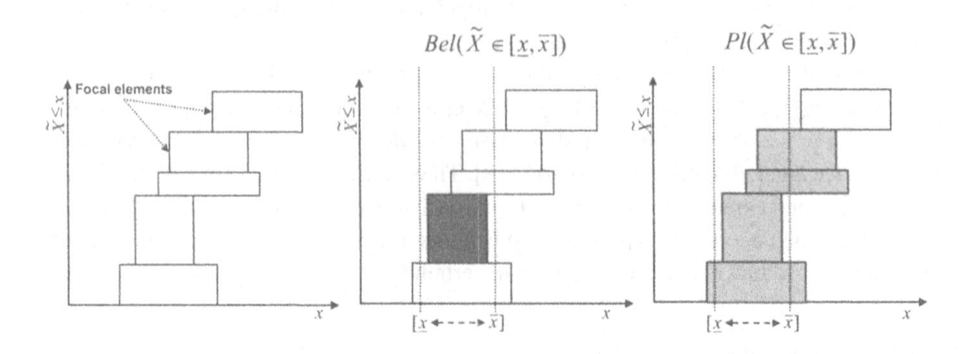

Fig. 3.3. Left: exemplary plot of a BPA. Center: Focal elements contributing to $Bel([\underline{x},\overline{x}])$. Right: Focal elements contributing to $Pl([\underline{x},\overline{x}])$. The height of a focal element represents its corresponding mass.

In case of continuous system functions (most system dependability models), the definition can be restricted to intervals $\overline{\underline{x}} = [\underline{x},\overline{x}]$ and thus $m : X^2 \longrightarrow [0,1]$ [57]. Similar to discrete probabilities, where it is possible to enumerate all values with non-zero probabilities, a BPA has a set of focal elements with non-zero masses.

Definition 3.5 (Focal element). *A focal element* $\overline{\underline{g}} := [\underline{g},\overline{g}]$ *is an interval with a non-zero mass* $m(\overline{\underline{g}}) > 0$.

Definition 3.6 (Random set). *The collection of all focal elements* $G := \{\overline{\underline{g_1}},\ldots,\overline{\underline{g_n}}\}$ *along with their masses is called a random set* $(G,M) := \{(\overline{\underline{g_1}},m(\overline{\underline{g_1}})),\ldots,(\overline{\underline{g_n}},m(\overline{\underline{g_1}}))\}$.

A BPA may be completely described by the set of focal elements and their associated masses. Throughout this thesis, the random set (G,M) or equivalently the BPA m will be used to describe a belief variable $\widetilde{X}$. A mass $m([\underline{x},\overline{x}])$ can be interpreted as the evidence of a "small best-worst-case scenario" expressed by an interval on a parameter $[\underline{x},\overline{x}]$. The best-case and worst-case values are represented by the interval bounds $\underline{x}$ and $\overline{x}$. Inside the interval $[\underline{x},\overline{x}]$, no distribution (e. g. uniform or triangular) is assumed. Figure 3.3 (left) shows a BPA describing $\widetilde{X}$. Focal elements are drawn as rectangles with their height representing their mass. It can be seen that, compared to probabilistic representations, a specific value of x could be included in several focal elements.

If another degree of freedom for uncertainty modeling is desired, the uncertainty model on x could be levered from a probabilistic to a DST representation. In DST, masses are assigned to intervals without distributing these masses further on points inside the interval. Therefore it is not possible to give an exact probability $P(\widetilde{X} \in \overline{\underline{x}})$ or even $P(\widetilde{X} = x)$ that the event under consideration is contained in interval $\overline{\underline{x}}$ or at value x. However, upper and lower bounds on these probabilities may be calculated. Associated with each BPA are two functions $Bel, Pl : X^2 \longrightarrow [0,1]$ which are referred to as belief and plausibility of an event.

Definition 3.7 (Belief, Plausibility). *Belief* $Bel(\widetilde{X} \in \overline{\underline{x}})$ *and plausibility* $Pl(\widetilde{X} \in \overline{\underline{x}})$ *of an interval* $\overline{\underline{x}} \subseteq X$ *are defined as:*

$$Bel(\widetilde{X} \in \underline{\overline{x}}) = \sum_{A \subseteq \underline{\overline{x}}} m(A) \tag{3.6}$$

$$Pl(\widetilde{X} \in \underline{\overline{x}}) = \sum_{A \subseteq X : A \cap \underline{\overline{x}} \neq \emptyset} m(A) \tag{3.7}$$

$Bel(\widetilde{X} \in \underline{\overline{x}}) \leq Pl(\widetilde{X} \in \underline{\overline{x}})$ because $A \neq \emptyset, A \subseteq \underline{\overline{x}} \Rightarrow A \cap \underline{\overline{x}} \neq \emptyset$ and thus, if $m(A)$ contributes to $Bel(\underline{\overline{x}})$, then it contributes to $Pl(\underline{\overline{x}})$, too. In figure 3.3 (center), it can be seen that only the mass of the blue focal element contributes to $Bel([\underline{x}, \overline{x}])$, because it is the only one which is completely enclosed by $[\underline{x}, \overline{x}]$. Regardless of the inner distribution of the focal element's mass, it will be included in $[\underline{x}, \overline{x}]$. Figure 3.3 (right) illustrates the plausibility of an event. The masses of all green focal elements contribute to $Pl([\underline{x}, \overline{x}])$ because of the nonempty intersection with $[\underline{x}, \overline{x}]$. It is possible to distribute the mass of each focal element on points, which are enclosed in $[\underline{x}, \overline{x}]$.

$Bel(\widetilde{X} \in \underline{\overline{x}})$ and $Pl(\widetilde{X} \in \underline{\overline{x}})$ can be interpreted as bounds on the probability $P(\widetilde{X} \in \underline{\overline{x}})$. Informally, the belief function represents the maximal value that the estimating person despite all epistemic uncertainty "believes" to be smaller than $P(\widetilde{X} \in \underline{\overline{x}})$, the plausibility function represents the highest "plausible" value of $P(\widetilde{X} \in \underline{\overline{x}})$. Both values can be interpreted as the best and worst-case on a probability. Belief and plausibility will be used to display a BPA. In analogy to probabilistic notation, $Bel/Pl(x) := Bel/Pl(\widetilde{X} \leq x)$.

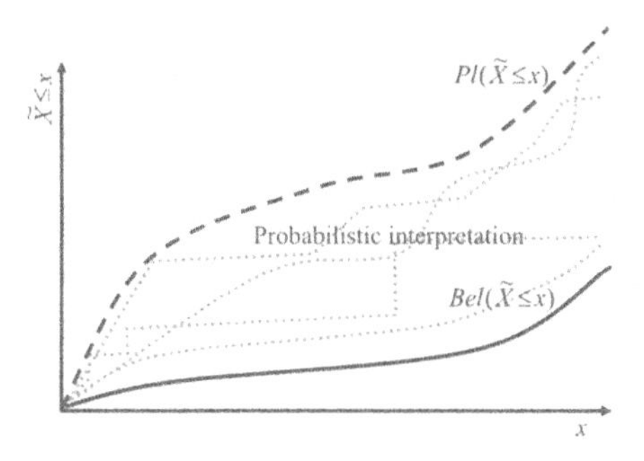

Fig. 3.4. Dempster-Shafer BPA (displayed: $Bel(\widetilde{X} \leq x), Pl(\widetilde{X} \leq x)$). Enclosed are some possible CDFs limited by belief and plausibility.

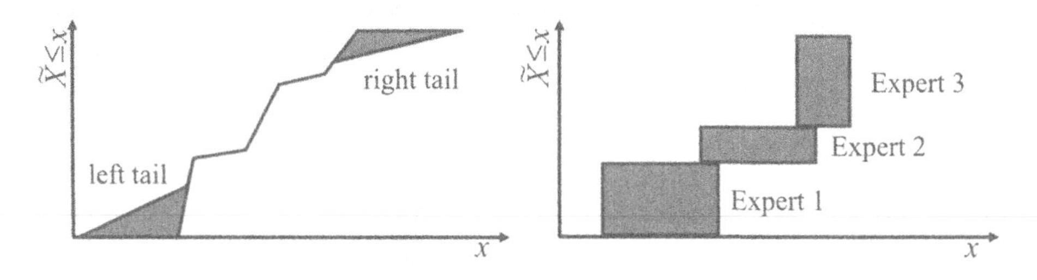

Fig. 3.5. Examples of uncertainty quantification in DST. Left: Quantifying distribution tails. Right: Aggregation of three best-worst-case estimates.

Figure 3.4 shows this probabilistic interpretation of a BPA. A common way to visualize the BPA on a variable is to plot $Bel(\widetilde{X} \leq x)$ and $Pl(\widetilde{X} \leq x)$ in analogy of a CDF. Included are some possible CDFs bounded by $Bel(\widetilde{X} \leq x)$ and $Pl(\widetilde{X} \leq x)$.

To ilustrate possible uses of the extended modeling flexibility, two cases of DST uncertainty quantification are shown (figure 3.5). The first case is a distribution on x, where the tail behavior is not well known. In a probabilistic approach, the tails can be estimated, but there will be uncertainty involved if this estimation is true. In a DST approach, the known part of the distribution may be defined, but the tails can be bounded without specifying the exact distribution. The second case shows the aggregation of three expert estimates on x. In DST there are several ways of merging probabilistic and interval estimates to a combined BPA (section 3.4). The distribution-free uncertainty represented by the intervals will be preserved in the aggregated BPA.

3.3.3 An Illustrative Example

Uncertainty modeling in DST covers some similar ideas but also some that are very different from probabilistic modeling. Throughout this chapter, an illustrative example will help to demonstrate the modeling process. The example is illustrated as simple as possible (for a realistic application, see chapter 7). The system is a reliable cabling with two redundant lines. The illustrative system function ϕ is represented by the fault tree in figure 3.6. The aim is to evaluate the probability that the system fails in its warranty period, estimated at 15,000h of operation. The failures of component "cable 1 broken" and component "cable 2 broken" are described by Weibull distributions $F_{Weib}(t,\alpha,\beta)$ with known shape parameter β, but unknown characteristic lifetime α. Let the characteristic lifetime of component 1 and component 2 be x_1 resp. x_2, then the system model is given by:

$$\phi(x) = \phi\begin{pmatrix} x_1 \\ x_2 \end{pmatrix} = F_{Weib}(15000, x_1, 2.3) \cdot F_{Weib}(15000, x_2, 2.1) \tag{3.8}$$

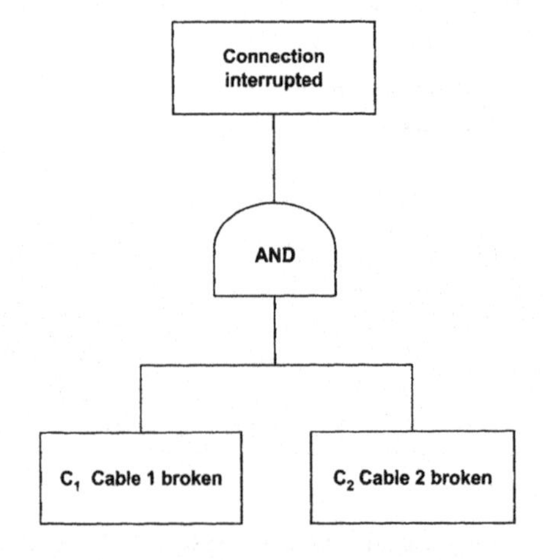

Fig. 3.6. Illustrative example: 2-Component fault tree

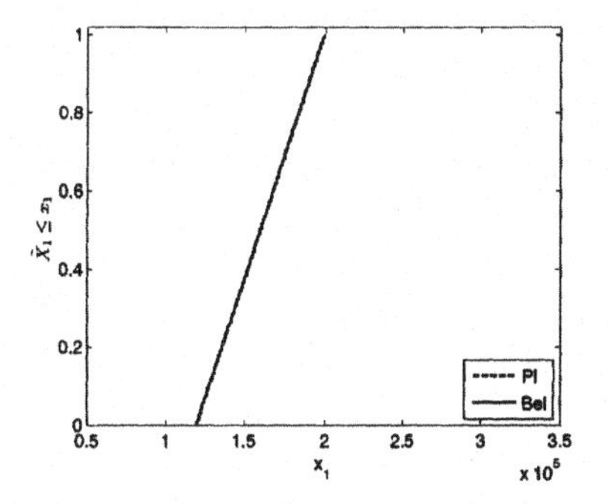

Fig. 3.7. Expert A's estimate on the char. lifetime of component 1, $m_{1,1}$ (200 samples).

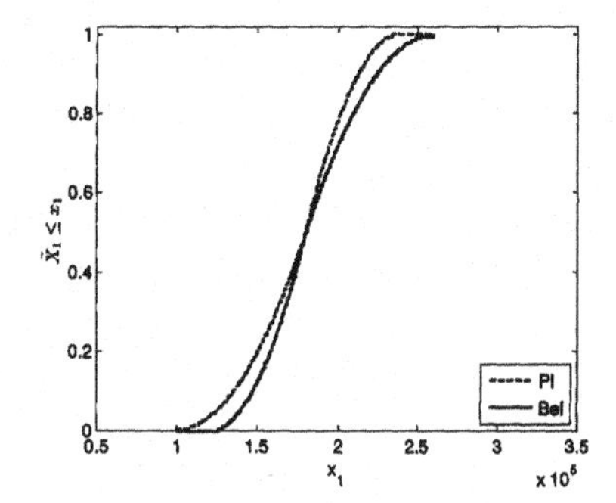

Fig. 3.8. Expert B's estimate on the char. lifetime of component 1, $m_{1,2}$ (200 samples).

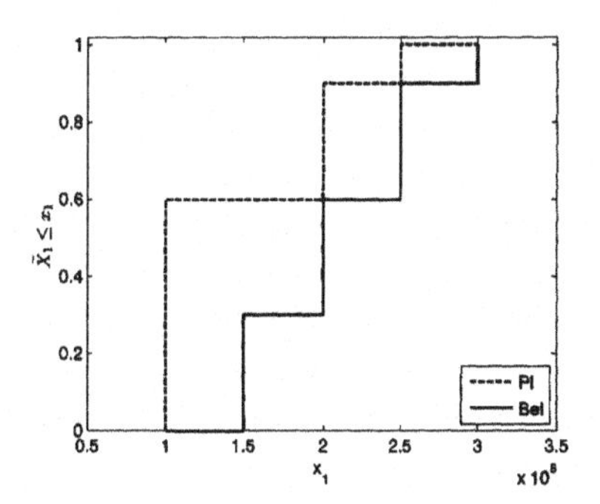

Fig. 3.9. Expert C's estimate on the char. lifetime of component 1, $m_{1,3}$.

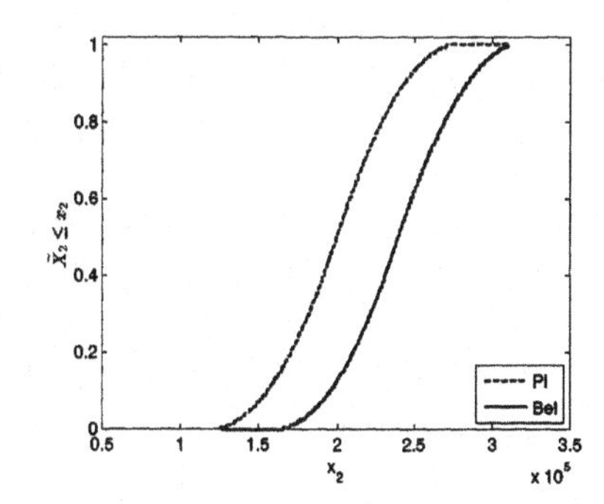

Fig. 3.10. Expert D's estimate on the char. lifetime of component 2, $m_{2,1}$.

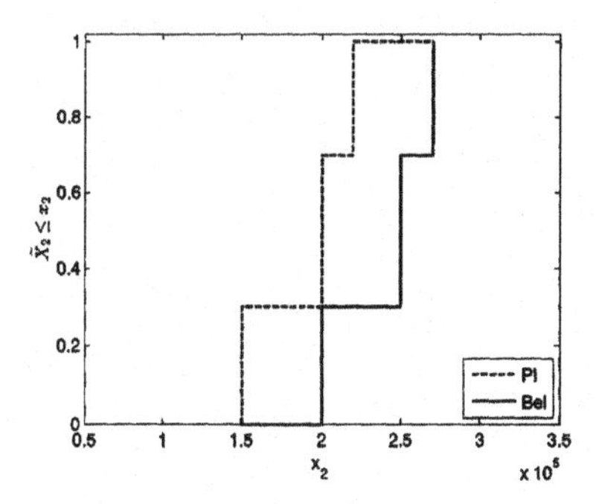

Fig. 3.11. Expert E's estimate on the char. lifetime of component 2, $m_{2,2}$.

If no further uncertainty is regarded, the characteristic lifetimes x_1 and x_2 are deterministic values and the system function $\phi(x)$ returns a system failure probability without uncertainty (e. g. $\phi(x)$=1E-5). In an early design stage, the vector x is not exactly known. Therefore in the example, three experts (A-C) are eager to provide estimates on the characteristic lifetimes. To illustrate the difference between probabilistic and DST representation of uncertainty, expert A gives a probabilistic estimate $m_{1,1}$ for component 1. He estimates that the characteristic lifetime of x_1 can be defined by a probabilistic variable that is uniformly distributed between 120,000h and 200,000h (figure 3.7).

Expert B gives an estimate $m_{1,2}$ on the BPA describing x_1. According to him, the char. lifetime of component 1 is triangular distributed with mode 180,000h. However he is in doubt about the right choice of the dispersion and gives only bounds on this quantity [60,000h, 80,000h] (figure 3.8). Both estimates are sampled to obtain discrete distributions / mass functions.

Expert C gives a set of intervals with their mass function $m_{1,3}$. He estimates four likely regions for the char. lifetime together with their likelihood. According to him, $m_{1,3}([100{,}000\text{h}, 150{,}000\text{h}]) = 0.3$, $m_{1,3}([100{,}000\text{h}, 200000\text{h}]) = 0.3$, $m_{1,3}([200{,}000\text{h}, 250{,}000\text{h}]) = 0.3$ and $m_{1,3}([250{,}000\text{h}, 300{,}000\text{h}]) = 0.1$. The resulting BPA is plotted in figure 3.9.

Component 2 is estimated by two experts (D, E). Expert D provides an estimate $m_{2,1}$ using a triangular distribution with uncertain mode [200,000h, 240,000h] but dispersion 80,000 h (figure 3.10). Expert E estimated failure intervals $m_{2,2}([150{,}000\text{h}, 200{,}000\text{h}]) = 0.3$, $m_{2,2}([200{,}000\text{h}, 250{,}000\text{h}]) = 0.4$ and $m_{2,2}([220{,}000\text{h}, 270{,}000\text{h}]) = 0.3$ (figure 3.11). Experts A-C have each provided their opinion on the char. lifetimes of component 1 and component 2 that need to be merged to one aggregated BPA. Various aggregation methods will be covered in the next section.

3.4 Aggregation

DST encompasses a great advantage in its natural ways of combining different sources of evidence. If there are k estimates on component i represented by the k BPAs $m_{i,1}, \ldots$, $m_{i,k}$, various methods have been proposed to construct an aggregation m_i of the BPAs. Comparing overviews are given in [47] and [149]. Perhaps the most popular method but also the most controversial part in DST is Dempster's rule [150]:

$$m_i(\underline{\bar{x}}) = \frac{\sum_{\underline{\bar{a}}_1 \cap \ldots \cap \underline{\bar{a}}_k = \underline{\bar{x}}} m_{i,1}(\underline{\bar{a}}_1) \cdot \ldots \cdot m_{i,k}(\underline{\bar{a}}_k)}{1 - K} \tag{3.9}$$

$$K := \sum_{\underline{\bar{a}}_1 \cap \ldots \cap \underline{\bar{a}}_k = \emptyset} m_{i,1}(\underline{\bar{a}}_1) \cdot \ldots \cdot m_{i,k}(\underline{\bar{a}}_k) \tag{3.10}$$

Dempster's rule presumes that each m_i encloses the real value with at least one focal element ($K \neq 1$). If $K = 1$, the results are unspecified. K is often referred to as the "conflict" between the mass functions. It is the sum of all mass combination without a common intersection. The masses of the resulting BPA are normalized by this value. If

$K = 1$, there is no value/interval that has a non-zero mass in all $m_{i,1}, \ldots, m_{i,k}$. The visual interpretation of $K = 1$ is that there is no value that all experts agree upon. Dempster's rule can be very time-consuming, the amount of focal elements in the resulting BPA is limited by $O(|G|^n)$ with n belief functions with a maximum of $|G|$ focal elements.

Dempster's rule can give counter-intuitive results, if the amount of conflict is large. Yager [181] has proposed an alternative rule, which is more robust in this case. Its difference to Dempster's rule is that the conflict K is not used for normalization. Instead, K is added to $m(X)$. Informally, this means that all belief which can't be further separated, is put on the whole frame of discernment X. This rule is expressed by the following equation:

$$m_i(\underline{\overline{x}}) = \begin{cases} \sum\limits_{\underline{\overline{a}}_1 \cap \ldots \cap \underline{\overline{a}}_k = \underline{\overline{x}}} m_{i,1}(\underline{\overline{a}}_1) \cdot \ldots \cdot m_{i,k}(\underline{\overline{a}}_k) & ,\text{if } \underline{\overline{x}} \neq X \\ m_{i,1}(X) \cdot \ldots \cdot m_{i,k}(X) + K & ,\text{if } \underline{\overline{x}} = X \end{cases} \tag{3.11}$$

The interpretation of K is that conflictual evidence is not negligible, but shows that some focal elements are incompatible. This incompatibility is interpreted as a support of X. A comparison between both rules in the area of safety can be found in [144].

Dempster's and Yager's rule require that each estimate contains at least one focal element enclosing the true value of the event ($K \neq 1$). Otherwise, Dempster's rule is undefined, and Yager's rule would contain only one focal element, $m(X) = 1$. In expert estimates this precondition is not necessarily met. Thus, [57] propose instead the method of weighted mixing as a robust alternative. Weighted mixing allows assigning a degree of importance $w_1, \ldots, w_k$ to specific estimates. If this is not desired, all weights are set to 1. The weighted mixture of the BPAs $m_{i,1}, \ldots, m_{i,k}$ assigns to each interval the averaged mass of all BPAs. It is given as:

$$m_i(\underline{\overline{x}}) = \frac{\sum\limits_{j=1}^{k} w_j m_{i,j}(\underline{\overline{x}})}{\sum\limits_{j=1}^{k} w_j} \tag{3.12}$$

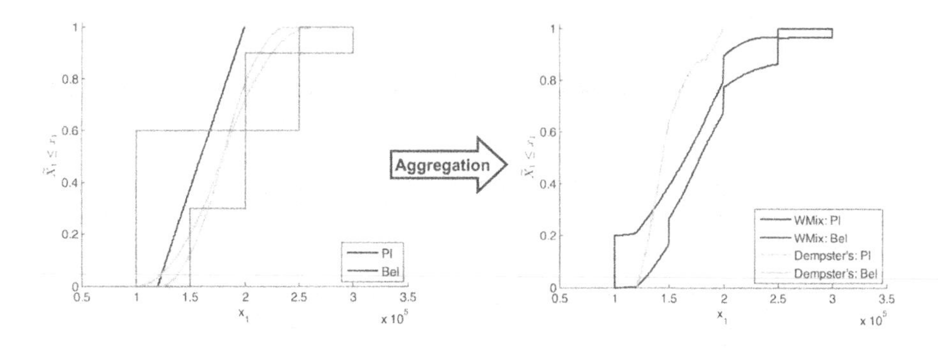

Fig. 3.12. Left: Estimates on component 1 provided by experts A-C. Right: Aggregation using both weighted mixing and Dempster's rule.

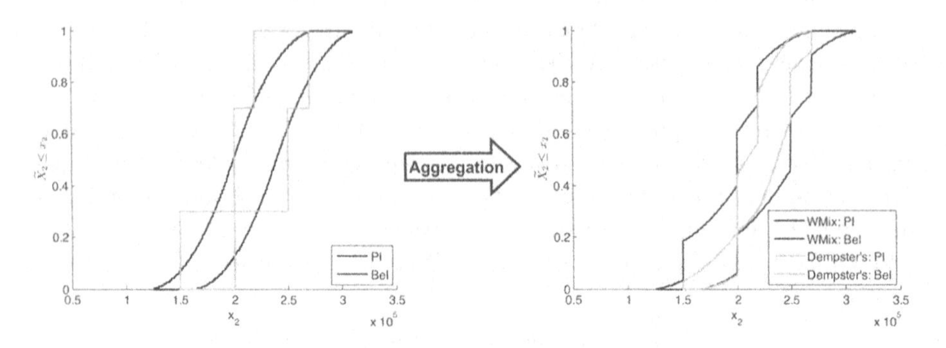

Fig. 3.13. Left: Estimates on component 2 provided by experts A & B. Right: Aggregation using both weighted mixing and Dempster's rule.

The amount of focal elements of a BPA merged by weighted mixing is limited by $O(|G| \cdot n)$ and thus considerably smaller than in Dempster's rule. Figure 3.12 shows the aggregation of expert A-C's estimates using a weighted mixture with equal weighting factors and Dempster's rule. While the resulting weighted mixture returns an averaged belief function, Dempster's rule is much less robust. As expert A provided only estimates on points, the resulting aggregation cannot contain any interval uncertainty, regardless of the other experts' estimates (due to the intersection term in eq. 3.9).

In figure 3.13, where expert B provided interval estimates starting at 150,000h and ending at 270,000h, the aggregation of Dempster's rule is limited to this area. The weighted mixture contains some belief outside these margins, too, resulting in an aggregation that seems to be more "adequate" to the human eye.

3.5 Dependency

Dependencies between input variables can have an important influence on the results of the uncertainty analysis. However they are often overlooked in the prediction process. This may be caused by various reasons: Either a possible causal relation was not recognized by the engineers. In this case, the dependency cannot be predicted, regardless of the chosen model. However, there are other reasons to assume independence. Dependencies can be overlooked because they do not seem important for component dependability analysis [55]. In this case, the dependability analyst can only hint to the possible influence of dependencies between input variables on the dependability prediction. The third reason is that dependencies are known or suspected, but their causal relation is too complex to be expressed. Thus, independence is assumed. This can be prevented, if dependencies between component properties may be estimated without revealing the complex causal relations. Hence, dependency modeling is a crucial part of the presented uncertainty model.

Dependency modeling covers a possible way to obtain the joint probability $\widehat{X}$ or the joint BPA $\widetilde{X}$ that describes the vector x from the marginal probabilities $\widehat{X}_1, ..., \widehat{X}_n$ resp. marginal BPAs $\widetilde{X}_1, ..., \widetilde{X}_n$. Joint and marginal probabilities can be defined as:

Definition 3.8 (Joint probability). *The joint probability is the probability of the co-occurrence of two or more events [177].*

Definition 3.9 (Marginal probability). *The marginal probability is the probability of one event $P(\widehat{X} \in A)$, regardless of another event [175].*

The common approach if forming a joint distribution is to assume independence between the marginal probabilities. Independence between events intuitively means that knowing whether event A occurs does not change the probability of B. In probability theory, independence can be defined as:

Definition 3.10 (Independence). *Any collection of events - possibly more than just two of them - are mutually independent if and only if for any finite subset $\widehat{X}_1 \in A_1, ..., \widehat{X}_n \in A_n$ of the collection [176]:*

$$P(\widehat{X}_1 \in A_1 \wedge ... \wedge \widehat{X}_n \in A_n) = P(\widehat{X}_1 \in A_1) \cdot ... \cdot P(\widehat{X}_n \in A_n) \tag{3.13}$$

Thus, two events are dependent, if this equivalence does not hold. Even if there is absolutely no knowledge about the type and strength of dependence between two events, bounds on the probability of the joint event exist. The so-called Fréchet boundaries for conjunction are these limiting probabilities:

$$P(\widehat{X}_1 \in A_1 \wedge ... \wedge \widehat{X}_n \in A_n) \geq \max\left(1 - n + \sum_{i=1}^{n} P(\widehat{X}_i \in A_i), 0\right) \tag{3.14}$$

$$P(\widehat{X}_1 \in A_1 \wedge ... \wedge \widehat{X}_n \in A_n) \leq \min\left(P(\widehat{X}_1 \in A_1), ..., P(\widehat{X}_n \in A_n)\right) \tag{3.15}$$

The Fréchet boundaries represent the two extreme cases of dependency (complete dependence, opposite dependence) and can theoretically be used to model total lack of knowledge on the dependency behavior. Because of the enormous amount of uncertainty introduced if hedging the dependencies with them, they are only useful as the most conservative modeling approach. In dependability prediction, at least the direction of the dependence (positive / negative) is known. Either components such as basic failure events tend to occur together (possible common causes) or inhibit each other. But it is difficult to imagine a case where absolutely nothing is known about the dependencies. Estimates on the dependency are important to be included in the uncertainty model, but should not artificially blow up the prediction uncertainty. Therefore, other dependency models are necessary.

3.5.1 The Concept of Copulas

In the probabilistic framework, copulas [154] are a way to introduce dependencies between several random variables into the uncertainty model. Copulas offer a way to specify the CDF of a joint distribution by knowing the CDFs of the margins and some information about their dependency relations. In other words, copulas are a way to "glue" marginal distribution functions $\widehat{X}_1, ..., \widehat{X}_n$ together to a joint distribution $\widehat{X}$. In system dependability prediction, the CDFs of the component failure distributions

$F_1(x_1),...,F_n(x_n)$ can be considered as margins, and the CDF of the joint distribution $F_{1,...,n}(x_1,...,x_n)$ may be used to obtain the joint probability distribution $m_{1,...,n}$ necessary for the prediction of the system failure probability. Copula functions give the probability of the joint distribution as:

Definition 3.11 (Copula). *An n-dimensional copula is a multivariate distribution function $C(u)$ with uniform distributed marginal distributions $u := (u_1,...,u_n), u_1 \in [0,1],...,u_n \in [0,1]$ and the following properties [123]:*

1. $C : [0,1]^n \rightarrow [0,1]$
2. *C is grounded:* $\exists i = 1,...,n : u_i = 0 \Rightarrow C(u) = 0$
3. *C has margins C_i which satisfy* $C(1,...,1,u_i,1,...,1) = u_i$
4. *C is n-increasing*

Given this definition, if $F_1(x_1),...,F_n(x_n)$ are CDF, $C(F_1(x_1),...,F_n(x_n))$ is a multivariate CDF with margins $F_1,...,F_n$. But it does not state that each multivariate CDF can be represented by margins and copula. Sklar's theorem [154] proves this and allows the use of copulas for dependency modeling as it makes the separation of both marginal distributions and dependencies possible.

Definition 3.12 (Sklar's Theorem). *Let $F_{1,...,n}$ be a an n-dimensional distribution function with margins $F_1,...,F_n$. Then $F_{1,...,n}$ has a copula representation:*

$$F_{1,...,n}(x_1,...,x_n) = C(F_1(x_1),...,F_n(x_n)) \tag{3.16}$$

If $F_1,...,F_n$ are continuous, C is unique. Otherwise C is uniquely determined on $Ran(F_1) \times ... \times Ran(F_n)$, where $Ran(F_i)$ denotes the range of F_i. In other words: an arbitrary joint distribution can be defined by knowing only the copula and the margins. The significant characteristic of probabilistic modeling with copulas is the separation of the component distributions (the margins) and the dependency representation. This yields the advantage that both can be estimated independently. Domain engineers can provide information on the component failure probabilities. System engineers later on predict possible dependencies between estimates without altering these estimates. A second benefit is that by specifying copula parameters, engineers predefine which estimates are correlated without needing to explicitly model reasons (such as the same type of component) leading to this dependency.

Copulas are mainly applied in financial risk prediction [131, 49], where model inputs are regularly known to be correlated by complex mechanisms not included in the model. The decoupling between margins and copulas allows the separate parametrization and eases the propagation through the system model [109]. Their usage is yet to become popular in dependability prediction. However, copulas can be a valuable tool for dependability prediction with little data [55].

3.5.2 Copula Types

Different copulas and copula families are in common use. Three special copulas were already introduced in section 3.5: maximal positive dependence, maximal negative dependence and independence. The extremal copulas, the Fréchet bounds C_+ respective C_- constrain the set of possible copulas:

$$C_-(F_1(x_1),...,F_n(x_n)) := \max\left(1-n+\sum_{i=1}^{n} F_i(x_i), 0\right) \tag{3.17}$$

$$C_+(F_1(x_1),...,F_n(x_n)) := \min(F_1(x_1),...,F_n(x_n)) \tag{3.18}$$

The one by far most frequently applied is the "unintentionally" used product copula C_P. It represents the independence assumption and is defined as:

$$C_P(F_1(x_1),...,F_n(x_n)) := \prod_{i=1}^{n} F_i(x_i) \tag{3.19}$$

The independence and the total dependence relations can therefore be seen as special cases of copula modeling. Figure 3.14 shows the copulas C_+, C_- and C_P for two dimensions. Plotted on the x-axis and y-axis are the marginal values, the z-axis gives the joint probability obtained by the copula. A visual interpretation of the Fréchet bounds C_- and C_+ is that each possible copula function (e. g. C_P) is completely located in the space that is spanned by these functions.

But copulas are only useful if they ease *specify* dependencies and thus add further degrees of freedom for the estimator. It is therefore necessary to use copulas which can be parametrized. Among the different copula families, the most popular copulas are the Archimedean copulas, the Gaussian copula and the t-copula [49]. All include C_-, C_+ and C_P as special cases. Examples for the family of Archimedean copulas are the Frank [61] and the Mardia copula [116]. Archimedean copulas are popular in case of two variables, but can't be readily extended to the multivariate case [113].

The most commonly used copula for more than two dimensions is the Gaussian copula. Its big advantage is the easy communicability. The set of parameters for estimating the dependencies is a $n \times n$ correlation matrix ρ. The Gaussian copula is defined as:

$$C_G(F_1(x_1),...,F_n(x_n),\rho) = F^{n,\rho}_{Gauss}(F^{-1}_{Gauss}(F_1(x_1)),...,F^{-1}_{Gauss}(F_n(x_n))) \tag{3.20}$$

F^{-1}_{Gauss} is an inverse standard normal distribution (mean zero, standard deviation 1) and $F^{n,\rho}_{Gauss}$ a multivariate standard normal distribution with correlation matrix ρ. The concept

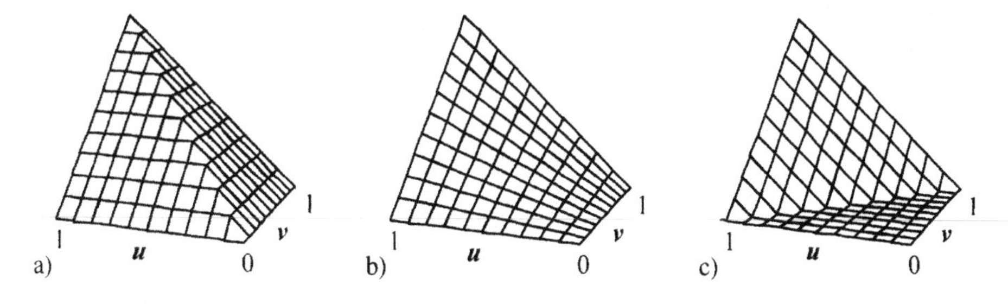

Fig. 3.14. Copula functions with margins u, v. a) Fréchet upper bound $C_+(u,v)$, b) Independence $C_P(u,v)$, c) Fréchet lower bound $C_-(u,v)$ [123].

of the Gaussian copula is to map the dependency structure onto a Gaussian multivariate distribution. The input marginal probabilities $F_1(x_1),...,F_n(x_n)$ are converted to values of a standard normal Gaussian distribution using F_{Gauss}^{-1}. In a second step, the cumulative probability is calculated by $F_{Gauss}^{n,\rho}$ using the dependency parameters ρ.

However it is necessary to mention that ρ is not the correlation matrix between the margins but the correlation of these margins transformed to a standard normal distribution. ρ is better understood as a matrix of dependency parameters with $\rho_{ij} \to 1$ indicating a growing positive dependency and $\rho_{ij} \to -1$ a growing negative dependency between components i and j.

3.5.3 Applying Copulas to Model Joint Imprecise Distributions

Copulas are a simplified, yet powerful way to model dependencies via marginal distributions. Thus, it is of interest to transfer this probabilistic concept of dependence to DST. Unfortunately, there is no such thing as a precise, marginal distribution in DST.

Nevertheless, the application of copulas may be extended to discrete Dempster-Shafer distributions as shown in [55]. As outlined, the equivalence in eq. 3.3 holds only for precise masses. For imprecise probabilities, $F(x) = P(\widetilde{X} \le x) \in [Bel(\widetilde{X} \le x), Pl(\widetilde{X} \le x)]$ is not perfectly known but bounded by intervals. $F_{1,...,n}(x) = C(F_1(x_1),...,F_n(x_n)$ is therefore interval-valued. So far, there is no problem in extending the concept. However, it is not possible to reconstruct a single, precise joint BPA from an imprecise joint CDF $F_{1,...,n}$. Further assumptions or redefinitions are needed. [55] propose to induce an order on the focal elements which is used to build a CDF-like margin. This is the direct transfer of probabilistic copulas to the imprecise case.

In the probabilistic case, a mass of the joint distribution $m_{1,...,n}(x_1,...,x_n)$ can only be non-zero if $\forall i = 1,...,n : m_i(x_i) > 0$. Hence, a CDF $F(x)$ is calculated by summing up all non-zero masses $m(x')$ with $x' \le x$ (eq. 3.3). The same holds for the DST case. Masses of the joint BPA $m_{1,...,n}(\overline{\underline{x}}_1,...,\overline{\underline{x}}_n)$ can only be non-zero if $\forall i = 1,...,n : m_i(\overline{\underline{x}}_i) > 0$ [180]. If copulas were to be applied in DST, it is first necessary to create an equivalent to a CDF. For each $i = 1,...,n$, a total order $<_i$ on the focal elements $\overline{\underline{g}}_{i,j} \in G_i$ is induced so that $\overline{\underline{g}}_{i,1} <_i \overline{\underline{g}}_{i,2} <_i ....$ Using this order it is possible to obtain a cumulative mass function $\breve{F}_i(\overline{\underline{g}}_{i,j})$ for all focal elements:

$$\breve{F}_i : G_i \longrightarrow [0,1] \tag{3.21}$$

$$\breve{F}_i(\overline{\underline{g}}_{i,j}) := \sum_{k=1}^{j} m_i(g_{i,k}) \tag{3.22}$$

With this "trick", it is now possible to generate a joint cumulative mass function $\breve{F}_{1,...,n}$ for all focal elements of the joint BPA via copulas:

$$\breve{F}_{1,...,n}(\overline{\underline{g}}_{1,j_1},...,\overline{\underline{g}}_{n,j_n}) = C(\breve{F}_1(\overline{\underline{g}}_{1,j_1}),...,F_n(\overline{\underline{g}}_{n,j_n})) \tag{3.23}$$

To illustrate the effects of dependency on the example, two dependency scenarios are compared. Scenario 1 assumes independence between the two components. The correlation matrix ρ_I for representing independence is the unit matrix. The second scenario

(ρ_C) assumes that there is a strong correlation between the characteristic lifetime estimates of both components. The engineers estimate the correlation matrices ρ_I and ρ_C as:

$$\rho_I = \begin{pmatrix} 1 & 0 \\ 0 & 1 \end{pmatrix} \text{ and } \rho_C = \begin{pmatrix} 1 & 0.7 \\ 0.7 & 1 \end{pmatrix} \tag{3.24}$$

This dependency model reflects the following behavior. If x_1 and x_2 are independent, then a high value of the characteristic lifetime of component one has no influence on the characteristic lifetime of component 2. In case of a correlation of 0.7, if one characteristic lifetime is small, then the other is very likely small, too. If components 1 and 2 are from the same manufacturer or are working under the same stress level, a dependency can be modeled.

3.6 Propagation through System Functions

In the propagation step, the component failure vector $\widetilde{X}$ must be propagated through the system function $\phi(\widetilde{X})$. The following notation is introduced: The joint BPA describing $\widetilde{X}$ is denoted as $m_{1,\ldots,n}$ with focal element vectors $\underline{\overline{g}} := (\underline{\overline{g_1}}, \ldots, \underline{\overline{g_n}})^T$. The BPA describing $\phi(\widetilde{X})$ is denoted by m_ϕ. Propagating a focal element through the system function is not as straightforward as the propagation in case of probabilistic functions. Being a synthesis between probabilistic and interval arithmetic, DST relies on optimization to propagate focal elements through the system function. m_ϕ is determined by [180]:

$$m_\phi(\underline{\overline{x}}) = \sum_{\underline{\overline{g}}:\phi(\underline{\overline{g}})=\underline{\overline{x}}} m_{1,\ldots,n} \begin{pmatrix} [\underline{g_1}, \overline{g_1}] \\ \vdots \\ [\underline{g_n}, \overline{g_n}] \end{pmatrix} \tag{3.25}$$

The amount of computation for an exact propagation is large. $O(|G_{1,\ldots,n}|) = O(|G_1| \cdot \ldots \cdot |G_n|)$ calculations of $\phi(\underline{\overline{g}})$ must be performed. The exponential growth with the number of parameters makes the exact solution only feasible if both n and $|G_1| \cdot \ldots \cdot |G_n|$ are

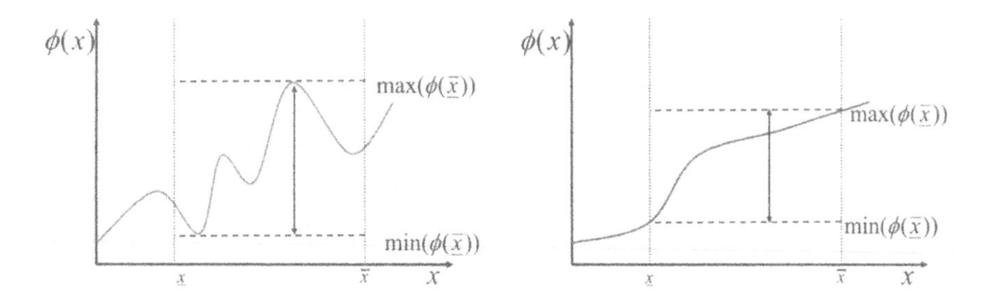

Fig. 3.15. Propagation in Dempster-Shafer theory. Bounds on $\phi(\underline{\overline{g}})$ depending on the shape of ϕ (1-D schematic representation). If ϕ is monotonous, the extrema are located at the borders of $\underline{\overline{g}}$.

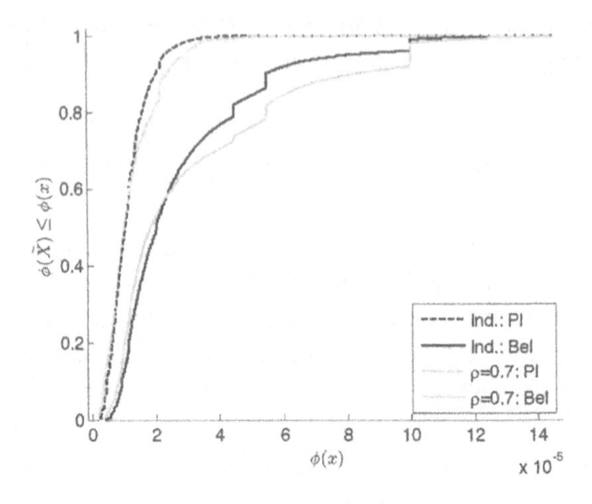

Fig. 3.16. System failure probability obtained after propagating the estimated component BPAs through the fault tree. Black: independent characteristic lifetimes. Green: characteristic lifetimes correlated with $\rho = 0.7$.

small. As in common probabilistic propagation, if $\widetilde{X}$ has a too large / infinite amount of focal elements, sampling techniques such as Monte Carlo sampling from $m_{1,\ldots,n}$ can be applied [117].

In eq. 3.25 the interval propagation $\phi(\overline{\underline{g}})$ occurs, which is solved as:

$$\phi(\overline{\underline{g}}) = \left[\min_{a \in \overline{\underline{g}}} \phi(a), \max_{a \in \overline{\underline{g}}} \phi(a)\right] \tag{3.26}$$

Thus, the propagation of a focal element involves the solution of two optimization problems ($\min \phi(a)$, $\max \phi(a)$). This equation holds only for continuous ϕ, to which this approach is restricted. For discontinuous functions (which are not common in dependability), the result would be a discontinuous set enclosed in the calculated interval.

Table 3.2. Propagation of variables in the illustrative example

$\overline{\underline{g}}$	$m_{1,2}(\overline{\underline{g}})$	$\phi(\overline{\underline{g}})$	$m_{1,2}(\phi(\overline{\underline{g}}))$
$[2.00E5, 2.50E5]$ $[2.2E5, 2.7E5]$	0.015	[0.0149,0.0343]	0.015
$[1.66E5, 1.66E5]$ $[2.28E5, 2.68E5]$	4.17E-6	[0.0355,0.0475]	4.17E-6
$[1.00E5, 1.50E5]$ $[2.20E5, 2.70E5]$	0.015	[0.0434,0.1461]	0.015
⋮	⋮	⋮	⋮

Table 3.3. Some characteristic values of the system failure probability

	Independence (ρ_I)	Correlated (ρ_C)
$E(p_s)$	[4.16E-5 8.98E-5]	[4.41E-5 9.92E-5]
$\mathrm{Med}(p_s)$	[3.83E-5 6.79E-5]	[3.72E-5 6.26E-5]
$Q_5(p_s)$	[1.49E-5 2.91E-5]	[1.35E-5 2.55E-5]
$Q_{95}(p_s)$	[8.02E-5 2.20E-4]	[9.62E-5 2.90E-4]
Bel/Pl(p_s¡1E-4)	0.02/0.30	0.04/0.34

Figure 3.15 illustrates the finding of the lower and upper bounds of an 1-D system function. On the left hand side, the bounds of $\phi(\underline{\overline{g}})$ must be obtained by optimization. The function of the right hand side is monotonous, representing a system dependability function. At the interval boundaries, $\phi(\underline{\overline{g}})$ has its maximal values. The optimization may be a bottleneck if the effort of solving the two optimization problems is high. Functions that tend to be (at least locally) easy to solve by optimization methods do not increase the calculation effort substantially. Fortunately, most reliability and safety problems are continuous and monotonously increasing [88]. Even in the largest block diagrams and fault trees, decreasing component failure probabilities do not reduce the system failure probability and thus the calculation can be simplified to:

$$\phi(\underline{\overline{g}}) = \left[\phi(\underline{g}), \phi(\overline{g})\right] \tag{3.27}$$

If ϕ is not as "well-behaved", optimization heuristics may be applied [3]. In [6], meta-models are used for a convenient speed of the uncertainty propagation.

The joint BPA of the illustrative example $\widetilde{X}$, obtained in section 3.5 can be propagated through ϕ by Monte Carlo sampling. In table 3.2, excerpts of the joint BPA and its propagation results are shown. The resulting plot of $\phi(\widetilde{X})$ for both scenarios (independent and correlated components) is shown in figure 3.16. A high component correlation especially influences the Bel/Pl values in very low and very high failure probability regions. The uncorrelated scenario performs similar to the correlated scenario. Table 3.3 shows some properties of $\phi(\widetilde{X})$. It can be seen that the bounds on the quantiles $Q_5(p_s)$ are higher for the uncorrelated scenario and for $Q_{95}(p_s)$ in the correlation scenario. The expected value of p_s, $E(p_s)$ and the median $Med(p_s)$ intervals are comparable.

If the illustrative example results were to be analyzed, it is especially interesting that component dependency influences $Q_{95}(p_s)$. If the system under investigation is a safety-critical system, the dependency between the estimates can play an important role. The belief and plausibility $Bel/Pl(p_s > 1E-4)$ that p_s exceeds $1E-4$ are given. While the belief is very low, there is a plausibility of 0.30 / 0.34 that this threshold is surpassed.

The results in table 3.2 can be interpreted as follows. If the exceedence of a threshold is tested, e. g. if it should be shown that $p_s > 1E-4$ will occur only with a low probability $P(p_s > 1E-4) < 0.1$, then this probability is bounded by the belief $Bel(p_s > 1E-4)$ and the plausibility $Pl(p_s > 1E-4)$. In the independence case, $Bel(p_s > 1E-4) = 0.02$ and $Bel(p_s > 1E-4) = 0.30$. The conclusion therefore is that it is (under the given amount of uncertainty) neither possible to show the exceedence nor the shortfall of the threshold.

After obtaining the results of the uncertainty study, they need to be interpreted and communicated. Here is a fundamental difference between DST and probabilistic modeling. DST allows quantifying uncertainty by interval width and by the distribution of focal elements. It is possible to calculate properties such as the expected value of a BPA similar to probability theory. The results are intervals, best- and worst-cases on the expected value.

One point needs to be mentioned, because it represents a trap in the application of DST: Averaging intervals of result properties to point values is questionable. In the quantification phase, uncertainties that could not be described by distributions were described by intervals, "best-worst-case" estimates. This interval uncertainty is propagated and preserved in the results. Averaging intervals would render the uncertainty quantification useless. This can be illustrated in comparison to a best-worst-case study. If the analyst tries to obtain the worst- and the best-case for an input parameter range, he would never average these values at the end but provide both quantities as results.

3.7 Measures of Uncertainty

Measuring the amount of uncertainty in a model is highly important to benefit from the extended uncertainty modeling. It can not only be seen where dependability problems emerge, but also, where there is still a high amount of uncertainty in the project. In section 3.8, sensitivity analysis methods for determining the contribution of component uncertainties to the system uncertainty are introduced. These methods require measures of uncertainty to evaluate the component contributions to these measures.

In DST, uncertainty measures split up into randomness measures and nonspecifity measures. Randomness is broadly speaking the distribution of the masses in the BPA and similar to uncertainty in probabilistic models. Nonspecifity on the other hand is uncertainty contributed by interval width. The wider the focal elements, the less specific is the resulting set of distributions.

Several possible generalizations of Shannon entropy [152] for DST were proposed. Not for all, such as for the aggregate uncertainty presented in [2], algorithms for real numbers are available [121, personal communication]. The fact that the width of the intervals has no direct influence on the uncertainty measure is an aspect which the randomness measures have in common. The Shannon entropy for probabilities is given as:

$$H(\widehat{X}) = -\sum_{x \in X} P(x=\widehat{X}) \log_2(P(x=\widehat{X})) \tag{3.28}$$

In the equation, $0 \cdot \log_2(0) = 0$ is defined. If extending this equation to Dempster-Shafer, $P(x=\widehat{X})$ needs to be replaced by an adequate measure. Proposed by Klir is the measure of dissonance DN:

$$DN(\widetilde{X}) = -\sum_{\underline{\overline{g}} \in G} m(\underline{\overline{g}}) \log_2(Pl(\underline{\overline{g}})) \tag{3.29}$$

If only focal elements on discrete values but not on larger sets exist, then $Pl(x=\widetilde{X}) = P(x=\widehat{X})$. Therefore, if applied to a probability distribution, DN is reduced to H. The

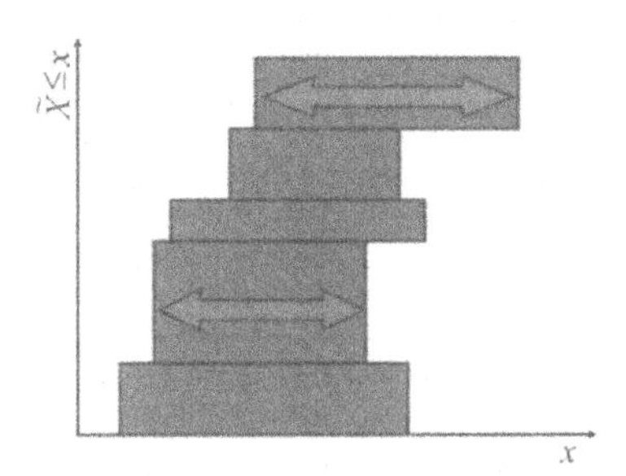

Fig. 3.17. Aggregated width measure *AW*. The larger the width of the focal elements, the higher the *AW* measure.

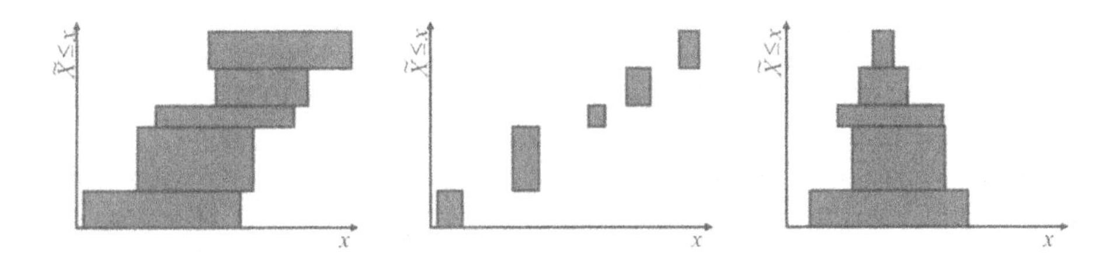

Fig. 3.18. Schematic representation of three BPAs. Left: BPA 1, high nonspecifity, low dissonance. Center: BPA 2, low nonspecifity, high dissonance. Right: BPA 3, medium nonspecifity, zero dissonance. All combinations of focal elements have a common subset.

interpretation of dissonance according to [86] is that *DN* is an entropy measure on the amount of conflict. This can be seen if eq. 3.29 is rewritten as:

$$DN(\widetilde{X}) = \sum_{\underline{\overline{g}} \in G} m(\underline{\overline{g}}) \log_2 \left(1 - \sum_{\underline{\overline{g}}' \in G : \underline{\overline{g}} \cap \underline{\overline{g}}' = \emptyset} m(\underline{\overline{g}}') \right) \tag{3.30}$$

The logarithmized term denotes the amount of conflict between $\underline{\overline{g}}$ and all other elements, the amount of focal elements that directly contradict $\underline{\overline{g}}$. *DN* is an entropy measure defined on this property.

Different nonspecifity measures for discrete frames of discernment *X* that resemble Shannon's entropy measure are proposed in [1] and [85]. For a discrete number of focal elements on the real line, [86] proposes to use a modified generalized Hartley measure:

$$GH(\widetilde{X}) = \sum_{\underline{\overline{g}} \in G} m(\underline{\overline{g}}) \log_2 (1 + \overline{g} - \underline{g}) \tag{3.31}$$

GH is an entropy measure on the width of all focal elements. Another way to estimate the nonspecifity is the aggregated width of all intervals[58]:

$$AW(\widetilde{X}) = \sum_{\underline{\overline{g}} \in G} m(\underline{\overline{g}}) (\overline{g} - \underline{g}) \tag{3.32}$$

Table 3.4. Uncertainty measures of component and system BPAs in the illustrative example (uncorrelated case)

System	GH	6.96E-5
	AW :	4.82E-5
	DN :	0.88
Component 1	GH	9.35
	AW :	2.39E4
	DN :	2.20
Component 2	GH	15.45
	AW :	4.50E4
	DN :	0.58

It measures the aggregated width of all intervals, which is the area between the *Bel* and *Pl* functions and thus is an intuitive uncertainty measure (figure 3.17). Both GH and AW depend on the scale of X, while DN is invariant to the choice of scale.

Parameter uncertainty can be reduced if both nonspecifity and dissonance decrease. Otherwise, it is possible to transform randomness to nonspecifity or vice versa. In figure 3.18, BPA 1 shows a high amount of nonspecifity. The focal elements are wide. However, the dissonance is low. Each focal element has intersections with several other focal elements, reducing the amount of conflict. BPA 2 has a very low amount of nonspecifity. Focal elements are very narrow. As there are no intersections, the degree of conflict for each focal element is rather large. BPA 3 shows a medium amount of nonspecifity but zero dissonance. These three examples help to better illustrate the interpretation of both measures. BPA 1 and BPA 2 are both expressing the enclosed uncertainty in different ways. The focal elements in BPA 1 are rather conservatively estimated. They are "nonspecific" in the sense that they cover a large range of values. On the other hand, it is thoroughly possible that several focal elements contain the real value of $\widetilde{X}$. In case of BPA 2, each focal element directly contradict all others. If each focal element was provided by one expert, then in BPA 1 all experts may be right, whereas in BPA 2 only one could be right.

From the results of the uncertainty measuring, different actions could be proposed. A high value of nonspecifity indicates that the focal elements are wide. Depending on the source of information, the reason might be underconfident experts that give excessively wide bounds (see section 3.1). A high amount of dissonance expresses a high amount of conflict regarding one variable. One possible reason may be that the experts have different understandings of the variable to estimate and the influences on this quantity. Another reason could lie in the overconfidence placed on the estimation when the real amount of information does not allow precise estimates. If different expert estimates are conflicting, discussions and iterative estimations may help either to find a common denominator or to enlarge the nonspecifity.

Table 3.4 shows the uncertainty measures in the illustrative example. The DN measure is larger for component 1, while the GH and the AW measures are larger for component 2. The system contains an average amount of dissonance (DN) compared to the components. The result itself may help to get a first impression on the amount of uncertainty in the model. But far more important is that they can be used as sensitivity measures for analyzing the components' uncertainty contribution. This method will be illustrated in the next section.

3.8 Sensitivity Analysis Using Uncertainty Measures

Sensitivity analysis is the process of varying model input parameters and observing the relative change in the model output. The main goal is to analyse the impact of the input variables on the quantities of interest. Sensitivity analyses are helpful to the decision-maker if he would like to refine the model or concentrate his analyses on certain parts of the modeling process. Components with high sensitivity require further information, as opposed to components with a relatively low impact.

First steps in the direction of sensitivity analysis in DST were made in [70] and [68]. An interesting specialty in DST is that it is not only possible to obtain sensitivity indices on the model output but also on the contribution to the different dimensions of uncertainty [58]. A sensitivity analysis regarding the nonspecifity can be carried out by measuring the nonspecifity of the whole model $GH(\phi(\widetilde{X}))$ / $AW(\phi(\widetilde{X}))$ in relation to the nonspecifity $GH(\phi(\widetilde{X}^i))$ / $AW(\phi(\widetilde{X}^i))$ of a "pinched" BPA $\widetilde{X}^i$:

$$s_{GH,i} = \left(1 - \frac{GH(\phi(\widetilde{X}))}{GH(\phi(\widetilde{X}^i))}\right) \tag{3.33}$$

$$s_{AW,i} = \left(1 - \frac{AW(\phi(\widetilde{X}))}{AW(\phi(\widetilde{X}^i))}\right) \tag{3.34}$$

$\widetilde{X}^i$ is the joint mass distribution assumed that $\widetilde{X}_i$ is perfectly known (i. e. replaced by a deterministic value). By sampling over various deterministic values enclosed by $\widetilde{X}_i$, a good approximation of $s_{AW,i}$ may be obtained. The sampling method applied is analogue to double-loop Monte Carlo sampling as applied in variance-based sensitivity analysis [141]. s_i reflects the contribution of $\widetilde{X}_i$ to the overall nonspecifity and can be used to analyze where additional component information has the largest impact on the result uncertainty.

The focus of the illustrative sensitivity analysis is to demonstrate how to study the effects of nonspecifity and dissonance of the two components on the system failure probability. Figure 3.19 shows the sensitivity analysis results on the illustrative example. It can be seen that, though both components contain a similar amount of uncertainty (table 3.4), their expected contribution to the overall system uncertainty is different. The contribution of component 1 to the system *GH* measure is slightly larger than the contribution of component 2. If the *GH* value of component 2 is reduced to zero, the expected reduction of the system *GH* measure would be about 31%. A reduction of the uncertainty of component 2 would lead to an expected reduction of the system *GH* measure by 28%. The *AW* measure shows stronger differences. While component 1 contributes about 43% of the uncertainty measured by *AW*, component 2 contributes only 23%.

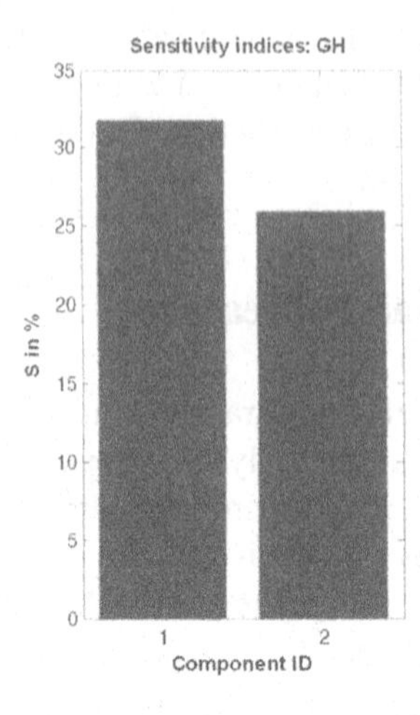

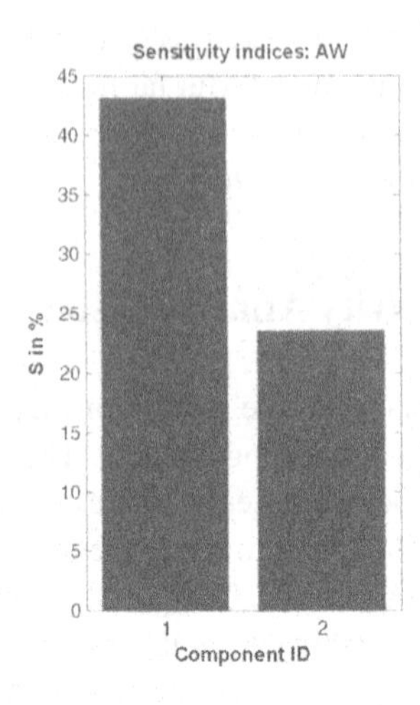

Fig. 3.19. Left: Sensitivity analysis on the GH measure (illustrative example). It can be seen that a slightly larger part of the uncertainty in the model is contributed by component 1. Right: Sensitivity analysis on the AW measure. Reduction of the uncertainty in component 1 has a strong influence on the result uncertainty. Component 2 plays only a minor role.

The sensitivity indices are in some way the essence of the uncertainty measuring process. It can directly be deduced on which components uncertainty reduction is especially valuable. Regarding the illustrative example, the sensitivity analysis shows clearly that component 1 should be the target of further tests or elicitation processes. Uncertainty reduction of component 2 could help to reduce the overall model uncertainty, but only to a smaller extent.

3.9 Comparing Dempster-Shafer Theory and Probabilistic Settings

In this thesis, DST is proposed as an alternative to probabilistic modeling. Therefore, it is necessary to discuss the applicability of DST in practice. The engineer in charge of an uncertainty study requires practical recommendations on when to and when not to use DST. There are two opposing "default" paradigms in quantifying uncertainty, the best-worst-case approach and the probabilistic one. DST allows using both paradigms in a combined representation and therefore in some situations it can be considered as an interesting alternative approach. By introducing interval uncertainty (e. g. by a wide focal element), a controversial or hardly-quantifiable source of uncertainty may be represented without the need to argue the choice of a distribution.

On the other hand, the advantages of a probabilistic approach compared to a best-worst-case scenario can also be utilized. Different best-worst-case scenarios can be weighted by masses. As DST is a generalization of the probabilistic representation, distributions, intervals and a mixture of both can be represented in the DST framework. This subsection tries to compare DST to both the deterministic and the probabilistic

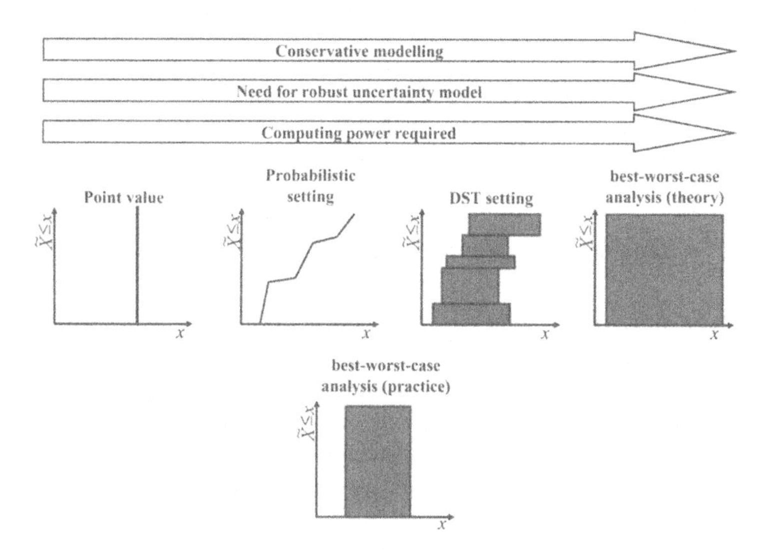

Fig. 3.20. Uncertainty settings arranged according to their conservativity, robustness and required computing power. This is a rough order as the real effort varies with the problem-specific uncertainty model and the system function.

settings from a practical point of view. As outlined, this comparison is partly based on practical experience and partly on research as DST is not commonly applied yet.

Figure 3.20 illustrates five uncertainty settings and their application depending on the requirements on the uncertainty study robustness. If the need for a robust uncertainty model is low, quite often a fixed point value is assumed (figure 3.20). This can be considered as a common approach in dependability, where failure probabilities are mainly modeled as deterministic values without uncertainty. If an uncertainty model is required, the practitioner may move to a probabilistic setting. The distribution describes uncertainty, however the analyst needs to specify the exact shape of the distribution. Sampling or other probabilistic propagation methods are required, raising the computational complexity of the analysis. DST is an appropriate choice if the need for a robust uncertainty model is even higher. Distributions may be enclosed within bounds, making the model more robust against a wrong choice of the distribution. The combination of sampling and optimization is more computationally intensive than purely probabilistic propagation. The most conservative approach would be theoretically the best-worst-case analysis supposing the whole parameter space could be accurately described. However, in practice rather a "pseudo" best-worst-case analysis is applied, with the largest *plausible* range of the parameter value used for obtaining the best- and worst-case, making the approach only maximally conservative respective to that bounds.

In DST, enhanced modeling freedom comes not for free. Along with the merge of interval uncertainty and probabilistic uncertainty models goes a rise in computational effort. Whether this rise is small or large depends on the shape of the system model (see Figure 3.15) and thus on the involved optimization task's complexity. If the model is complex and computational costs constitute a constraint, there are two adjustment screws in DST. The first is the number of samples of the joint BPA. The other is the accuracy of the optimization algorithm involved. The use of a meta-model replacing the full system model would of course come as an additional way out, with the same limitations as within probabilistic propagation.

3.9.1 The Decision between Dempster-Shafer Theory and Probability

This subsection aims to give some suggestions regarding the situations in which DST can be applied. A drawback in DST is that it is not as well standardized and as widely spread as probabilistic modeling. Therefore, the analyst planning to use DST may face some problems finding adequate methods and tools for his purpose. Nevertheless, some circumstances exist under which the application of DST could be valuable. In this section, several aspects that can influence a practitioners decision between DST and probabilistic settings are listed.

As discussed in section 3.2, an uncertainty study can aim at several goals and may be influenced by various factors. If it is expected that some of the assumptions taken for the uncertainty study may be doubted by a critical observer, DST can be applied to support some of the assumptions. A common target for possible criticism is the choice of the correct distribution. With DST, a whole range of distributions may be bounded without preferring one to the other. Therefore a DST study can be thought of as one way to reduce doubts regarding the uncertainty quantification.

The same holds if the analyst is not sure how to quantify the sources of uncertainty or if there is a very sensitive variable that is critical to be defined by a distribution while for others, convenient probabilistic representations exist. In this case, the analyst may refer to a best-worst-case estimate (or a weighted best-worst-case estimate such as a BPA), thus levering the whole study to a DST setting. If there exists some sort of "deterministic tradition" in a sense that the target readers would rather accept a deterministic best-worst-case analysis than a probabilistic setting, but the analyst has good reasons to prefer a probabilistic setting, DST can be used as an intermediate way between both paradigms.

In DST, an aggregation of expert estimates which are given in probabilistic, interval-valued or in a combined form is possible. DST has some natural ways of fusing data and expert estimates that are able to deal with these different representations (section 3.4). The resulting BPA still contains the characteristics of interval and probabilistic uncertainty. DST has the advantage / disadvantage of the interval paradigm. In its consequence it means that if the compliance of a threshold such as "failure probability $p_s > 6E-7$ with probability below 5%" is tested, the outcome could be:

- $Pl(p_s > 1E-4) < 0.1$: Compliance with threshold criterion despite all interval uncertainties.
- $Bel(p_s > 1E-4) > 0.1$: Violation of the threshold criterion despite all interval uncertainties.
- $Bel(p_s > 1E-4) < 0.1$, but $Pl(p_s > 1E-4) > 0.1$: It cannot be concluded if the criterion is reached or not, because the amount of interval uncertainty is too large.

Therefore DST maintains more than the probabilistic setting the fact that the amount of epistemic uncertainty in a model is large. Demonstrating compliance with a threshold in DST therefore may be a very supportive argument. On the other hand, a lot of discipline is necessary, because an excessive, pessimistic use of interval uncertainty may lead even faster to results that are ambiguous. If a probabilistic, deterministic or a DST approach is the optimal choice therefore depends heavily on the stakeholders and motivations of an uncertainty study.

4 Predicting Dependability Characteristics by Similarity Estimates – A Regression Approach

The presumption of similarity estimation is that companies do not solely create independent, completely innovative and new products. Product families with a number of variants are much more common. If products are re-used, dependability knowledge should be re-used as well. If already acquired data is efficiently included into dependability predictions of new products, then this might lead to a drastically improved accuracy/-cost ratio for dependability prediction. These thoughts spurred the development of the similarity estimation process described.

In the earliest development phases in particular, knowledge from similar components is a tremendously important source of evidence. Dependability figures of in-service components may be considered as very precise compared to numerical expert estimates on the dependability parameters. Instead of directly estimating, similarity measures between new and existing components can be elicited and thus both sources of evidence can be combined. Perhaps the most prosperous sources are dependability figures from inter-company data bases [142]. However, most companies treat their dependability data quite confidentially. Without the benefits of maintaining a shared data pool, there is no incentive to change this opinion. However, should the value of a shared data pool for internal dependability analyses rise, then this position may change. Efficient methods for the use of similar data can therefore present a good motivation to be less restrictive with the own dependability data.

This chapter presents a new way of dependability prediction by similarity estimation (figure 4.1). The starting point is a new component C_X without any known dependability data. The presumption of the prediction method is that there is data of similar components C_A, C_B, C_C available, e. g. failure distributions F_A, F_B, F_C. The target is to predict the failure distribution F_X for C_X. Two different machine-learning methods (neural networks, Gaussian processes) are used to learn the prediction task from training data and thus to support the engineer with the quantification of EDS dependability properties.

The chapter is structured as follows: In section 4.1, the transformation factor, a related concept proposed in [95] is introduced. Section 4.2 presents the expert elicitation procedure for the similarity prediction method and describes sources of uncertainty that arise with the elicitation. In section 4.3, the key of learning similarity prediction, the formulation of the regression problem to solve, is presented. Neural networks and Gaussian processes, two machine learning methods, are used to solve this regression

P. Limbourg: Dependability Modelling under Uncertainty, SCI 148, pp. 53–76, 2008.
springerlink.com

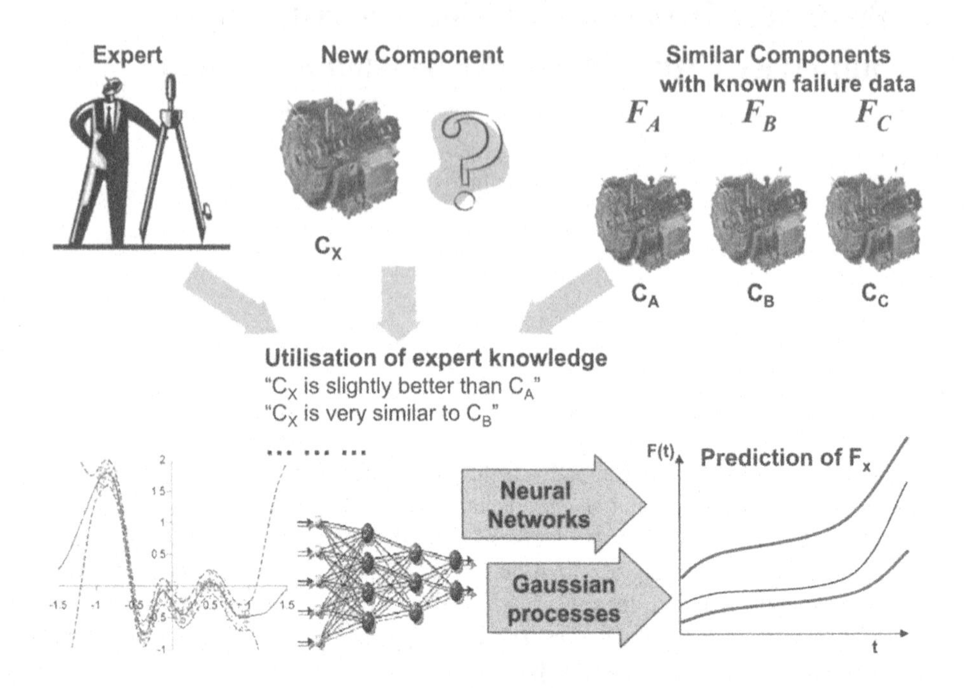

Fig. 4.1. The similarity estimation process

problem (section 4.4). To evaluate the proposed prediction methods, a scalable test suite is created and a real world prediction example is presented (section 4.5). In sections 4.6 and 4.7, neural networks and Gaussian processes are evaluated on their prediction capability using these test sets and conclusions on the usability of the prediction method are drawn.

4.1 Related Work: The Transformation Factor

There exists limited amount of work which combines the inclusion of similar component knowledge and expert estimates into dependability prediction; one approach which counts amongst the existing works is the transformation factor method. The concept of the transformation factor [95, 94] originates from the field of reliability demonstration. Considering a fixed point in time t_0, the initial situation of reliability demonstration is to demonstrate that, considering all uncertainties, $R(t_0) > p_c$ for a reliability value $p_c \in [0,1]$ with a confidence of $P(R(t_0) > p_c) \geq p_d$. A Bayesian framework is used to assess the probability density function $f_{Pri}(R(t_0))$. The prior probability of f_{Pri} is considered to be a beta distribution $F_{Bet}(A_1, A_2)$ specified by the parameters A_1 (number of components alive at t_0) and A_2 (number of components failed at t_0). This results in the following expression for f_{Pri}:

$$f_{Pri} : [0,1] \mapsto [0,1] \tag{4.1}$$

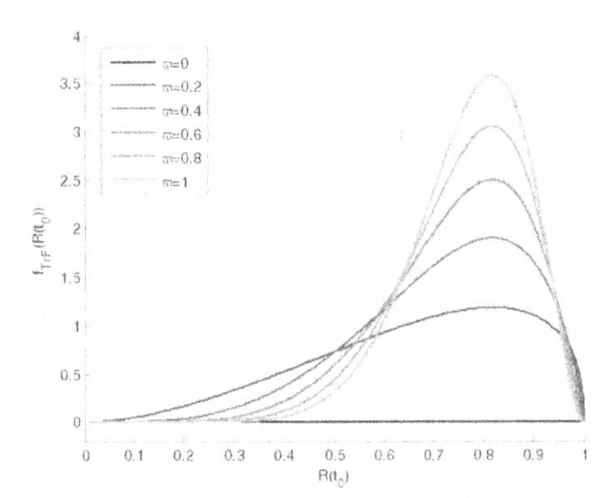

Fig. 4.2. Prior distribution of $R(t_0)$ for different values of ϖ with $A_1 = 10, A_2 = 3$. It can be observed how the shape of the PDF varies depending on ϖ.

$$f_{Pri}(R(t_0)) = \frac{R(t_0)^{A_1-1}(1-R(t_0))^{A_2-1}}{F_{Bet}(A_1,A_2)} \tag{4.2}$$

It is mathematically convenient to model the uncertainty around $R(t_0)$ with a beta distribution because its domain is equal to the range of $R(t_0) = [0,1]$. In a normal reliability demonstration study, the values A_1 and A_2 are determined by tests of the component of interest. The transformation factor approach comes into play if A_1 and A_2 are not directly assessed but originate from similar products. For a new product, reliability information of preceding products may not fit perfectly. The relevance of the old data can be adjusted by a real-valued qualifier $\varpi \in [0,1]$, the transformation factor. This factor resembles the "distance" qualifier introduced in section 4.2. The transformation factor denotes the grade of similarity between old and new component designs. The prior degrades to a completely noninformative one if the transformation factor decreases to zero:

$$f_{TrF}(R(t_0)) = \frac{R(t_0)^{\varpi(A_1-1)}(1-R(t_0))^{\varpi(A_2-1)}}{F_{Bet}(\varpi(A_1-1)+1, \varpi(A_2-1)+1)} \tag{4.3}$$

If $\varpi \to 0$, f_{TrF} converges to a uniform distribution $F_{Uni}(0,1)$, because the distribution $F_{Bet}(1,1)$ degenerates to an uniform distribution. If $\varpi \to 1$, the distribution converges to eq. 4.2. Figure 4.2 shows a prior distribution for different transformation factors. It can be observed how the PDF describing $R(t_0)$ varies depending on ϖ. For $\varpi = 1$, the PDF is very specific, while for $\varpi = 0.2$, it is almost uniform.

The transformation factor itself is only a mathematical representation to include similarity data and control its influence on the reliability estimate by the factor ϖ. In [73], a strategy to assess the transformation factor with a fuzzy model is presented. The transformation factor approach contains some fundamental differences compared to the method presented in the following sections. Firstly, it is necessary to know the sample size to express the uncertainty in the estimation. It is not possible to implement the model if the reliability value is known without information on the underlying trials (A_1,A_2). Secondly, it is only possible to quantify the relevance of old data for a new component, but not qualitative elements such as "the new product will be better". And thirdly, it relies on other algorithms to adjust the transformation factor, thus shifting the parametrization task one level higher. In contrast to these methods, the presented approach aims to *learn* similarity prediction. Using data from a test set generator as well as an application example, similarity estimation is formulated as a regression problem solved by neural networks and Gaussian processes. With this machine-learning approach, the adaptivity is increased. This allows for tailoring the prediction method to arbitrary dependability data sets. Thus, special requirements of the domain, company and project may be accommodated.

4.2 Estimation Procedure

The object of interest is a new component C_X in an EDS, without any (or little) knowledge about its dependability. However, it is presumed that a dependability data base is accessible that contains one or several similar components C_A, C_B, C_C with functions F_A, F_B, F_C obtained from dependability data. This may be test data or warranty figures from similar in-service components or older expert estimates. F_A, F_B, F_C may either be CDFs or failure rates. If C_X is not a completely new component but an existing component in a new system, test data from preceding systems may be available. One constraint that is not imposed is the restriction to specific distribution models. The dependability figures available may have an arbitrary form and fitting the data to a closed form distribution (e. g. Weibull) may unnecessarily bias the data. The prediction method therefore should be able to process empirical distributions.

4.2.1 Elicitation

The expert estimation process is separated into two steps: the selection of a set of similar components, and the estimation of different qualifiers that describe the grade of similarity.

4.2.1.1 Selection of One or More Similar Components

In this step, the expert selects similar components from the dependability data base. In an ideal case, a subset of similar components should be selected for the elicitation from an available large data base of components with dependability data. This leads to the question of what defines a "similar component". Stressing the special properties of the chosen machine-learning approach, it is not necessary to predefine a specific meaning on similarity other than: "A component is similar to the unknown component if the expert can estimate some commonalities/differences between their dependability behavior". Engineers are much better than expert systems in choosing components which they consider to share some properties with C_X. These properties may include similar dependability characteristics, shape, manufacturing process, the same charge or simply the impression that they are similar without any concrete reason.

To illustrate this meaning, three transmission systems C_A, C_B, C_C in the market and one new system C_X are assumed. An expert may choose transmission system C_X to be similar to C_A because they are produced in the same production period. Additionally he may choose C_B because of a usage profile similar to C_X. Or he may add C_C because it shares some building parts with C_X. As can be seen in this example, there is intentionally no fixed definition for "similarity" for not restraining the expert in his choice.

4.2.1.2 Estimation of Similarity Relations

The crucial step in the elicitation procedure is the estimation of similarity relations between the new component C_X and one of the selected components C_A. In the case study, three qualifiers are chosen which enable the experts to estimate the relation between components.

Distance. The distance qualifier describes how distinctly the components C_A and C_X differ from one another. If the distance is large, the expert assumes the relation between both components is only weak. If the distance is very low, the components are almost identical. Again, the interpretation of "difference" is left to the expert (e. g. the number of identical parts).

Quality. The quality qualifier provides an estimate on the quality difference between C_A and C_X. If the quality is low, C_X is assumed to have worse dependability values than C_A. If the quality is high, C_X is expected to perform better.

Wear out. The wear out qualifier describes how strong the new component will degrade comparatively at the end of its lifetime. If the value is high, the new component may degrade faster than the old component. In case of a low value, the new component, even if showing similar dependability characteristics at the beginning of the lifetime, will degrade slower.

All qualifiers are expressed by real values, and thus the relation between an old and a new component can be defined by the vector of qualifiers $q := (q_{dist}, q_{qual}, q_{wo})^T$. Experts express their belief in how strong these qualifiers describe this relation by providing an estimate as:

Distance. $q_{dist} \in [-1, 1]$: completely different $\leftrightarrow$ almost the same
Quality. $q_{qual} \in [-1, 1]$: much worse $\leftrightarrow$ much better
Wear out. $q_{wo} \in [-1, 1]$: much lower $\leftrightarrow$ much higher

The method is not restricted to this particular set of qualifiers. Other qualifiers or nomenclatures (such as infant mortality and environmental stress) are possible as long as they stay the same for the prediction and training process. In the example, qualifiers were directly provided in a real-valued form. However, a verbal representation could be implemented for the ease of use:

- "C_X is very similar to C_A"
- "C_X is moderately better than C_A"
- "C_X shows higher wear out than C_A"

By mapping verbal expressions to the numeric input (e. g. by table lookup), a verbal interface could be implemented.

The crucial assumption underlying the prediction algorithm is that if a new component C_X resembles a known component C_A, then F_X and F_A will also be similar in some way. If there is at least a small dependence, this dependence can be learned and used for prediction purposes.

The elicitation method as presented has several practical advantages. The expert does not have to know something about F_X but only something about C_X which makes it a true prediction method for the earliest development phases. This encapsulation of dependability prediction in a black box shows a second advantage: Domain experts who, albeit with commendable excellence in their domain, may not have good knowledge in dependability engineering, do not need to deeply understand dependability concepts and terms such as "failure distribution" or "survival probability". If failure probabilities are estimated directly, this understanding is required.

Each completed estimate contains one reference component C_A with known failure distributions C_A and an estimate for each qualifier value. It may be convenient to estimate relations to more than one component. The table in figure 4.3 shows possible outcomes of an estimation process. It is now to investigate, if a function f_{sim} can be constructed, which in some way takes q and F_A as inputs and predicts F_X.

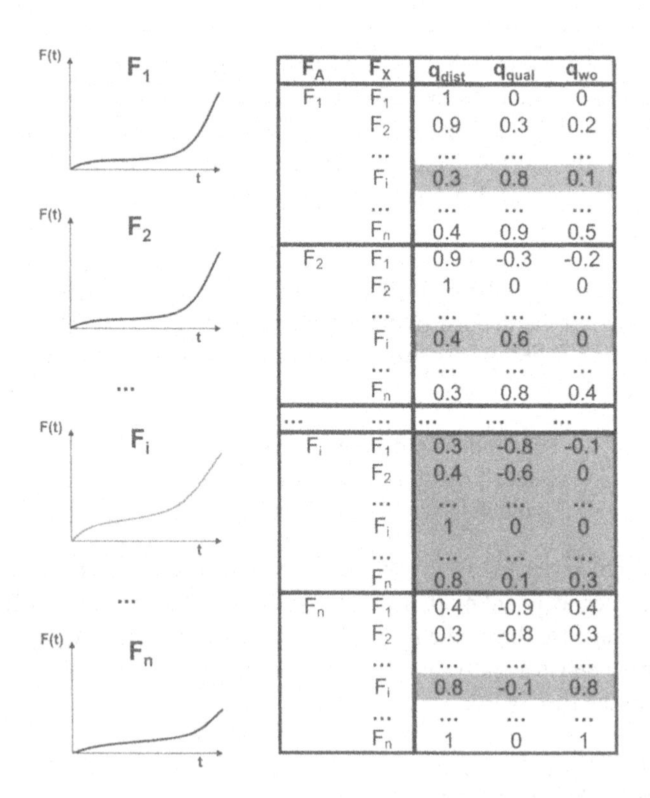

F_A	F_X	q_{dist}	q_{qual}	q_{wo}
F_1	F_1	1	0	0
	F_2	0.9	0.3	0.2
	...	...	...	...
	F_i	0.3	0.8	0.1
	...	...	...	...
	F_n	0.4	0.9	0.5
F_2	F_1	0.9	-0.3	-0.2
	F_2	1	0	0
	...	...	...	...
	F_i	0.4	0.6	0
	...	...	...	...
	F_n	0.3	0.8	0.4
...	...	...	...	...
F_i	F_1	0.3	-0.8	-0.1
	F_2	0.4	-0.6	0
	...	...	...	...
	F_i	1	0	0
	...	...	...	...
	F_n	0.8	0.1	0.3
F_n	F_1	0.4	-0.9	0.4
	F_2	0.3	-0.8	0.3
	...	...	...	...
	F_i	0.8	-0.1	0.8
	...	...	...	...
	F_n	1	0	1

Fig. 4.3. Elements of the test set. Components $C_1, \ldots, C_n$ with failure functions $F_1 \ldots F_n$ and information on the qualifier values of each function pair. F_i and its corresponding qualifiers are not revealed in the training process and solely used for testing purposes.

4.2.1.3 Providing Training Data

Aimed at defining a flexible approach that is adaptable to the specific requirements of the application domain and not a fixed prediction algorithm, machine-learning methods are utilized to learn from prediction examples. These learning methods can generalize from training data and will be used to infer f_{sim}, the function for predicting F_X. For this purpose, a set of predictions need to be presented as learning examples (figure 4.3). This training set contains a number of components $C_1, ..., C_n$ with known failure functions $F_1, ..., F_n$. Experts estimate the qualifiers for a number of combinations of failure functions $F_A \in \{F_1, ..., F_n\}$ and $F_X \in \{F_1, ..., F_n\}$. In the figure, F_i and its corresponding qualifiers (gray values) are not revealed in the training process. Instead, they are used for evaluating the approach.

With the prediction output F_X known, it is now possible to train a machine-learning method to mimic f_{sim}. The machine-learning method should learn to generalize from these examples how to predict F_X from F_A. In an iterative fitting process, the prediction is refined until the training cases can be predicted satisfiable. The trained prediction method is then applied to new estimation tasks with unknown output. If the prediction works well, it can be investigated by excluding some components during the learning phase. The prediction quality can be demonstrated by testing if these functions may be predicted using similarity estimates.

4.2.2 Inherent Sources of Prediction Uncertainty

The function f_{sim} is inherently uncertain. This uncertainty is introduced by several factors, mainly *ambiguity* and *limited relevance* (see section 3.1). *Information of limited relevance* rises the uncertainty, if the data of the component C_A at hand is not very similar to the component of interest C_X. *Ambiguity* is introduced by linguistic imprecision and insufficient specification of the predicted function. The meaning of qualifiers such as "slightly better" or "very similar" is not fixed and may describe sets of possible functions. But even if numerical qualifier values are estimated, the function F_X is insufficiently described. This is not because of inaccurate prediction methods but simply because there are many "slightly better" F_X for one given F_A. The method therefore can only try to predict bounds or probabilities on this estimate.

Among different experts, this qualifier meaning may even vary much stronger (uncertainty due to *conflicting beliefs*), and thus the sole prediction of the mean would be of no great use. It is of utter importance that the uncertainty is included in the model. Component dependability figures are used in dependability prediction of large systems. If the component plays a key role or is simply used in high numbers, small deviations in the predicted distribution may have enormous influences on system dependability figures. Therefore it is indispensable that these uncertainties are represented in the prediction output.

Beside these inherent sources, further uncertainties are introduced by the prediction method itself. The prediction model may not fit, the parametrization may be inadequate or the training data is simply not representing the real world. In this approach, these additional uncertainties will be reduced by carefully designing the network and evaluating both real and realistic artificial data.

4.3 Formulating Similarity Prediction as a Regression Problem

If a set of training data is available, the similarity prediction may be acquired by a machine-learning approach. Therefore, the prediction problem must be formulated in a way that it fits in the format necessary to apply common regression techniques. This section explains briefly the basics of regression before showing how similarity prediction can be represented as a function that can be regressed.

4.3.1 Regression

The basic form of a regression model can be formulated as a function

$$f_{reg} : X \times W \mapsto Y \tag{4.4}$$

$$y = f_{reg}(x, w) \tag{4.5}$$

which takes as arguments the input vector $x \in X$ and a parameter vector $w \in W$. The value y is the regressed / predicted output which is the object of interest. Both neural networks (section 4.4.1) and Gaussian processes (section 4.4.2) can be described in this general form as particular choices of f_{reg}. In neural networks, w represents the weight vector, while in Gaussian processes, w contains the parameters of the covariance function. The supervised learning (or training) process aims to adjust w to fit f_{reg} to the training set $(\psi_1;\xi_1)...(\psi_n;\xi_n)$. Each training pattern consists of an output $\xi_i \in Y$ which was generated by the unknown function to be learned from an input $\psi_i \in X$. The abbreviation used in the further text will be $\Psi := (\psi_1, ..., \psi_n)$ for the vector of all training inputs and $\Xi := (\xi_1, ..., \xi_n)$ for the vector of all training outputs. In [16] learning is defined based on this formalization as:

Definition 4.1 (Learning / Training). *Learning is the determination of w on the basis of the data set $(\Psi;\Xi)$.*

A machine-learning approach contains a particular model f_{reg} parametrized by w but also an optimization criterion which allows optimization algorithms to fit w to the training data. The optimization problem to solve is commonly the minimization of an error function $Err(w)$:

$$\min_{w \in W} Err(w) \tag{4.6}$$

By minimizing the error function on the training set, the model is fitted on the training data. The most common error function is the sum squared error function, measuring the deviation between predicted outputs $\check{\xi}_i := f_{reg}(\psi_i, w)$ and real outputs ξ_i for all patterns ψ_i:

$$Err(w) = \frac{1}{2} \sum_{i=1}^{n} (\check{\xi}_i - \xi_i)^2 \tag{4.7}$$

In the following it is shown how similarity prediction problem can reformulated in this framework. The resulting function can be learned by an arbitrary machine learning method.

Two different types of regression methods for solving this prediction function are investigated. Section 4.4.1 covers neural networks, a well-known and popular regression method applied in numerous areas. Gaussian processes are tested as a concurrent approach. This probabilistic regression model (also known as Kriging) has some special properties that are helpful to obtain estimates on the prediction uncertainty. Gaussian processes are introduced in 4.4.2.

4.3.2 Implementing the Regression Problem

Having carried out the estimation process, it is possible to describe the set of estimated qualifiers describing the relations between two components C_A and C_X with a vector $q = (q_{dist}, q_{qual}, q_{wo})^T$. For the creation of an adaptive prediction system, it is necessary to define the function that is to be regressed, a vector-valued function f_{sim} which can be learned from examples. Unfortunately the original regression problem to be solved is not vector valued, because the objective is to predict a whole function F_X, not only a single value. Furthermore, as discussed in section 4.2.2, the prediction itself needs to respect sources of uncertainty, which are inherent in the estimation procedure and cannot be reduced by training. Hence, similarity prediction needs to be reformulated in a way that it is vector-valued and respecting the aforementioned requirements.

As a simplifying assumption similar to the Markov property (the probabilities of future states depend only on the current state) is introduced: The function value of the unknown function $F_X(t_0)$ depends only on the time t_0, the function value $F_A(t_0)$ of C_A at time t_0, and q, the qualifiers describing the relation between C_X and C_A, but on no other value $F_A(t \neq t_0)$ of F_A. Then the regression problem can be simplified by processing one by one all points in time of the input function. For a given point in time, $F_X(t)$ can then be obtained via the similarity prediction function f_{sim} defined as:

$$f_{sim} : X \subseteq \mathbb{R}^5 \rightarrow Y \subseteq \mathbb{R} \tag{4.8}$$

$$F_X(t) = f_{sim}(x) = f_{sim}\begin{pmatrix} t \\ F_A(t) \\ q \end{pmatrix} \tag{4.9}$$

F_X is obtained from F_A by applying the prediction method on the points in time for which F_A is known while keeping q constant.

4.4 Learning Similarity Prediction

There are various ways of tackling the imposed prediction problem. Possible approaches can be divided into the "black box" and "gray box" methods. The former are very flexible, with a large set of parameters. If nothing or little about the underlying function is known, black box methods may be the right choice. The neural network regression investigated in section 4.4.1 and in [111] belongs to that area. Gray box methods impose strong restrictions and premises on the underlying function. In exchange, they manage with a very small amount of parameters. Learning processes may be reduced to simple optimization problems and exact solutions for the parameters may be found. A representative of the second group is Gaussian process learning, which is evaluated as a concurrent approach in section 4.4.2 and [107].

4.4.1 Neural Networks

Neural networks (NN) have proven their efficiency on regression problems over several decades and are still be considered as the tool of choice by many researchers. In dependability prediction, they are a popular tool for various regression tasks, e. g. [25]. Neural networks are well-known and many good overviews are existing [16, 19]. Therefore, in this thesis, a detailed introduction is omitted and only the changes needed to adapt them to similarity prediction are presented.

4.4.1.1 Input and Output Representation

Neural networks are used in various types of regressions. However, it is mainly the goal to predict deterministic functions. In the presented learning scenario, the object of interest is an imprecise function. It is known that there is irreducible uncertainty in the estimate that must be predicted. For this special case, the shape of the network can be modified for predicting imprecise, probabilistic models from precise training data. The signature of the function $f_{NN}(x,w)$ that represent the network is given as:

$$f_{NN}(x,w) = \begin{pmatrix} \check{F}_X \\ \overline{F_X} \\ \underline{F_X} \end{pmatrix} \tag{4.10}$$

Figure 4.4 visualizes the network. f_{NN} first scales the function using linear and logarithmic scaling and the network is trained in this scaled domain. Scaling makes the

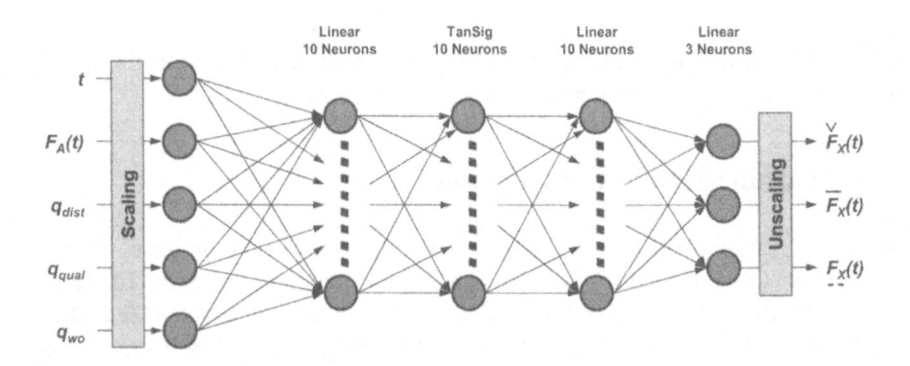

Fig. 4.4. Selected network design. The input is scaled, propagated through the layers (Linear, TanSig, Linear, Linear). Three output neurons predict the uncertain function $F_X(t)$.

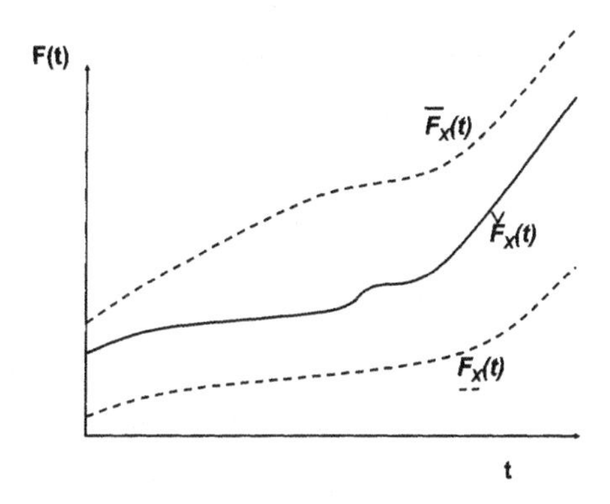

Fig. 4.5. Prediction of the function and its uncertainty bounds using neural networks

problem more accessible for the network, leading to better prediction results. For the prediction task, the network outputs representing the prediction results are unscaled to the original domain.

To reflect the uncertainty, f_{NN} has three outputs. While the first network output tries to predict $\check{F}_X$, the best estimate of F_X with respect to the inherent uncertainty, the second and third neurons predict upper ($\overline{F_X}$) and lower ($\underline{F_X}$) bounds on the function (figure 4.5). Thus the prediction results including both best estimate and surrounding uncertainties become useful for robust system dependability prediction in an EDS.

4.4.1.2 Customized Error Function

Neural network training tries to optimize the layer weights to minimize the error between the training patterns $\xi_1, \ldots \xi_n$ and their predictions $\check{\xi}_1, \ldots \check{\xi}_n$. Most methods rely on the backpropagation of an error vector through the net. As ξ are scalar values while $\check{\xi}$ are 3-dimensional vectors, training f_{NN} is therefore not possible without further modification of the network learning process.

For training a network with three outputs, a three-dimensional error criterion is mandatory. To be able to predict $\overline{F_X}$ and $\underline{F_X}$, the network training has to use a new error criterion that allows the second and third neuron to be trained for the uncertainty bounds. The error criterion should fulfill two functions in one. If the values to predict are found outside the bounds, the error value for the specific bound should be high. However, bounds should be very tight themselves to reduce imprecision. The applied error criterion $e(\xi_i, \check{\xi}_i)$ consists of three elements.

$$e(\xi_i, \check{\xi}_i) := \begin{pmatrix} e_1(\xi_i, \check{\xi}_{i,1}) \\ e_2(\xi_i, \check{\xi}_{i,2}) \\ e_3(\xi_i, \check{\xi}_{i,3}) \end{pmatrix} \tag{4.11}$$

The function e_1 represents the distance between the predictions of the first neuron $\check{\xi}_{i,1}$ and the correct output:

$$e_1(\xi_i, \check{\xi}_{i,1}) := \xi_i - \check{\xi}_{i,1} \tag{4.12}$$

e_2 and e_3 depend on the distance between the predictions of $\check{\xi}_{i,2}$ / $\check{\xi}_{i,3}$ and the correct output ξ_i. In contrast to e_1, this distance is weighted by its rank among all distances of the whole training set:

$$e_2(\xi_i, \check{\xi}_{i,2}) := (\xi_i - \check{\xi}_{i,2}) \cdot Q(\xi_i - \check{\xi}_{i,2})^2 \tag{4.13}$$

$$e_3(\xi_i, \check{\xi}_{i,3}) := (\xi_i - \check{\xi}_{i,3}) \cdot (1 - Q(\xi_i - \check{\xi}_{i,3}))^2 \tag{4.14}$$

The value $Q(\xi_i - \check{\xi}_{i,j}) \in [0,1]$ returns the quantile that $(\xi_i - \check{\xi}_{i,j})$ obtains in the whole error sample $\{(\xi_1 - \check{\xi}_{1,j}), ..., (\xi_n - \check{\xi}_{n,j})\}$. $Q(\xi_i - \check{\xi}_{i,j}) = 0$ holds for the smallest (i. e. most negative) distance, $Q(\xi_i - \check{\xi}_{i,j}) = 1$ for the largest and $Q(\xi_i - \check{\xi}_{i,j}) = 0.5$ for the median distance. By squaring this quantile, the error vector is biased in the direction of the extremal errors. e_2 enlarges the highest positive errors, thus driving the prediction in this direction. e_3 enlarges the largest errors in the negative direction, driving the output in the direction of the lowest extremal predictions. This nonparametric approach was chosen because it is independent of the underlying distribution determining the error vector.

4.4.1.3 Network Design

The design of the network was an iterative process including several test series (e. g. documented in [107] and [111]). The best results were achieved using a network with four layers including linear and hyperbolic tangent sigmoid (TanSig) activation functions which is shown in figure 4.4. It was observed that the tangent sigmoid layer is important for "smoothing" the prediction and reducing the sensitivity to random noise. Larger networks led to overfitting in the tests, smaller network were less accurate.

The network weight optimization was done using gradient descent with momentum and adaptive learning rate backpropagation [66] as provided by MATLAB's neural network toolbox. It turned out to be much more efficient for the customized error function than the quasi-Newton algorithms or the Levenberg-Marquardt algorithm [67], proposed in [160] as the fastest for medium-sized networks.

4.4.2 Gaussian Processes

Gaussian processes [115, 145] are an alternative for the inference of $f_{sim}(x)$. Gaussian process learning (GP) assume that the training data $(\Psi; \Xi)$ is drawn from a function $f_{sim}(x) + \varepsilon$ where ε is an additional random noise factor. Thus, prediction uncertainty is included in the model. The goal is to infer $y = f_{sim}(x)$ for an unseen input x as a probability distribution $P(y|\Psi, \Xi, x)$. In contrast to the neural networks, where additional modifications for uncertainty estimation were necessary, this probabilistic approach allows to extract both best estimate and uncertainty bounds analytically. The basic idea in GP is to interpret the training set as a realization of a stochastic process.

Definition 4.2 (Gaussian process/GP). *Let $f(x), x \in X$, be a stochastic process. Then $f(x)$ is a Gaussian process if for any $x_1, ..., x_n \in X, n \in \mathbb{N}$, the vector $(f(x_1), ..., f(x_n))$ is a multivariate Gaussian distribution [115].*

A Gaussian process makes strong assumptions both about the noise and the interconnection between different function values which are not exactly fulfilled by the prediction problem. Nevertheless the results seem to mimic reality quite good. Applying definition 4.2, the approach assumes that the joint probability distribution of any n output values such as the training patterns $\Xi = (\xi_1, ..., \xi_n)$ is an n -dimensional Gaussian distribution with covariance matrix Cov_n:

$$P(\Xi|\Psi, Cov_n) = \frac{1}{Z_n} \exp\left(-\frac{1}{2}\Xi^T Cov_n{}^{-1}\Xi\right) \tag{4.15}$$

The construction of Cov_n and Cov_{n+1} hides the real learning process and will be explained later, Z_n is a normalization constant. Now the aim is to predict y using the trained model and the new pattern x. Applying Bayes' law results in:

$$P(y|\Psi, \Xi, x, Cov_{n+1}) = \frac{P(\Xi, y|\Psi, x, Cov_{n+1})}{P(\Xi|\Psi, x, Cov_{n+1})} = \frac{P(\Xi, y|\Psi, x, Cov_{n+1})}{P(\Xi|\Psi, Cov_n)} \tag{4.16}$$

The last expression holds because of the conditional independence between Ξ and x. Inserting eq. 4.15 into eq. 4.16 results in:

$$P(y|\Psi, \Xi, x, Cov_{n+1}) = \frac{Z_n}{Z_{n+1}} \exp\left(-\frac{1}{2}y^T Cov_{n+1}{}^{-1}y + \frac{1}{2}\Xi^T Cov_{n+1}{}^{-1}\Xi\right) \tag{4.17}$$

According to definition 4.2, sampling over different values of y results in a one-dimensional Gaussian distribution on y. A more efficient way is to compute the mean and variance of this distribution analytically [135]. Introducing the notations κ and k for parts of Cov_{n+1} as seen in eq. 4.18, mean function $\mu(x)$ and a variance function $\sigma(x)^2$ are given by:

$$Cov_{n+1} = \begin{pmatrix} Cov_n & k \\ k^T & \kappa \end{pmatrix} \tag{4.18}$$

$$\mu(x) = k^T Cov_n{}^{-1}\Xi \tag{4.19}$$

$$\sigma(x)^2 = \kappa - k^T Cov_n{}^{-1}k \tag{4.20}$$

$\mu(x)$ and $\sigma(x)^2$ will be used for computing the prediction intervals. Thus, the model is set up, but what are the parameters w to fit in a learning process? The only open point remaining is how to obtain the covariance matrix entries. This is done by a covariance function $f_{Cov}(\psi, \psi')$ proposed in [115] returning the covariance between two training patterns ψ, ψ':

$$f_{Cov}(\psi, \psi') = w_{I+1} \exp\left(-\frac{1}{2}\sum_{i=1}^{I} \frac{(\psi_i - \psi_i')^2}{w_i^2}\right) + w_{I+2} \tag{4.21}$$

where I is the input dimension (for similarity prediction $I = 5$). The parameter vector $w = (w_1, ..., w_{I+2})$ is fixed by an optimization algorithm to minimize the model error. w_{I+1} and w_{I+2} are overall scaling terms.

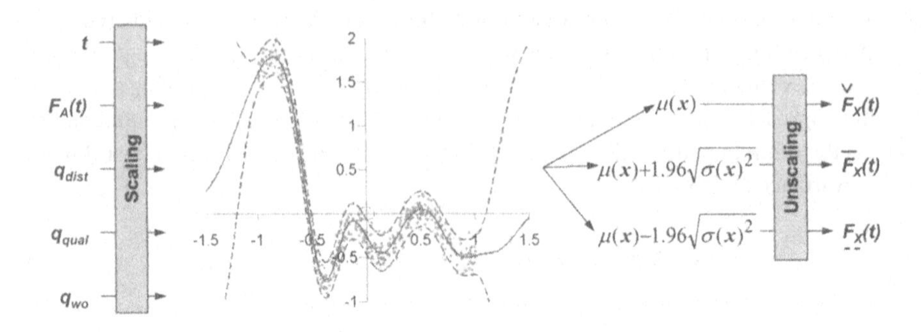

Fig. 4.6. Visualization of the Gaussian processes method for similarity estimation. The input is scaled and processed by the GP. From the mean function $\mu(x)$ and the variance function $\sigma(x)^2$, F_X is predicted.

If comparing this approach to the neural network approach shown in section 4.4.1, the gray box property can be observed. In a neural network, the number of free parameters is large. In GP, this number decreases enormously (in the example to 7). The interval boundaries are computed from the 95% confidence intervals of the normal distribution which corresponds to $\mu(x) \pm 1.96\sqrt{\sigma(x)^2}$. Therefore, the function f_{GP} for learning similarity prediction is given as:

$$f_{GP}(x,w) = \begin{pmatrix} \mu(x) \\ \mu(x) + 1.96\sqrt{\sigma(x)^2} \\ \mu(x) - 1.96\sqrt{\sigma(x)^2} \end{pmatrix} \tag{4.22}$$

To gain better learning results, linear and logarithmic scaling were used for both input and output of the function. After unscaling, the function values for F_X can be obtained. In figure 4.6, the GP approach is visualized. For the implementation of the GP approach, the spider package [173] for MATLAB was used.

4.5 Test Sets

To investigate the prediction capability of the presented approaches, two different prediction scenarios were tested. The first test set was generated by a newly developed test set generator which allowed an arbitrary number of test cases to be created. With this large data set, it is possible to evaluate and enhance the performance of the methods. The second test set is based on real failure data obtained from a family of mechatronic clutch systems provided by ZF Friedrichshafen AG.

4.5.1 Scalable Test Suite

There are several arguments for generating a scalable test set. The main reasons are a flexible amount of training data and full control on the generated functions. An arbitrary

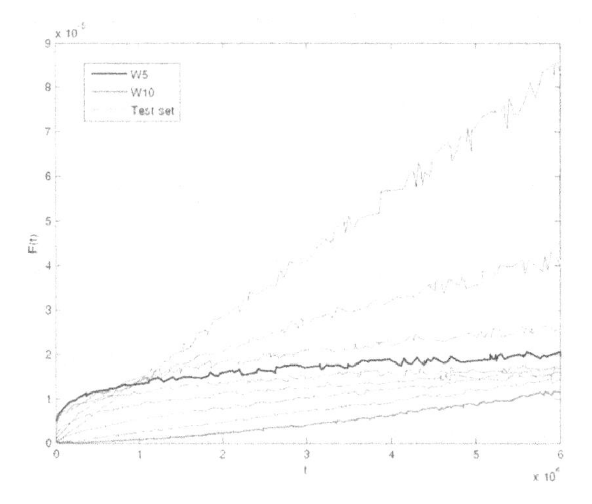

Fig. 4.7. Failure rates of an artificial test set. It can be seen that the components follow a Weibull distribution with strong noise. The highlighted functions W5 and W10 will be predicted in the tests.

number of failure rates may be generated, allowing the method to prove its efficiency on the basis of a large set of data with known properties.

The functions $F_1,...,F_n$, serving as training examples were generated from Weibull failure rates and obscured with random noise. The Weibull distribution is the common model to fit real failure data of mechanical components [130]. Its failure rate $\lambda_{Weib}(t)$ is given by:

$$\lambda_{Weib}(t) = \frac{\beta}{\alpha}\left(\frac{t}{\alpha}\right)^{\beta-1} \tag{4.23}$$

In [15] and [174], characteristic lifetimes and shape parameters are listed. The shape parameter β lies usually in an interval $\beta \in [1.1, 2.5]$ for components under normal wear out conditions. The characteristic lifetime α depends much stronger on the type of component under investigation. [174] e. g. gives $\alpha = 40,000$ as typical for valves. Boundaries for the test set parameters were derived from these values. α_{min} and α_{max} are bounds specifying the minimal and maximal α value, β_{min} and β_{max} limit the β values. The bounds on the Weibull parameters for the tests executed in this thesis were set to $\alpha_{min} = 30,000$, $\alpha_{max} = 100,000$ and $\beta_{min} = 1.2$, $\beta_{max} = 2.5$. The characteristic lifetime α_i of a function F_i is generated linearly by:

$$\alpha_i = \alpha_{min} + \frac{(i-1)(\alpha_{max} - \alpha_{min})}{n_F - 1} \tag{4.24}$$

The shape parameter $\beta \in [\beta_{min}, \beta_{max}]$ is a parabolic function with

$$\beta_i = \beta_{min} + (\beta_{max} - \beta_{min})\frac{((n_F + 1)/2 - i)^2}{((n_F - 1)/2)^2} \tag{4.25}$$

The test set applied contains $n_F = 10$ functions named W1,..., W10. Figure 4.7 shows several of the generated functions in one diagram (highlighted are the functions W5 and W10 that will be predicted in the tests). The introduced random noise leads to the ragged shape of the functions.

Each combination of two components (C_i, C_j) was rated with all three qualifiers. Qualifiers are subjective factors estimated by engineers and therefore difficult to fit into generator rules. The proposed systematics however may be justified by the specific properties of the Weibull distribution.

The distance qualifier $q_{dist,i,j}$ for a component combination (C_i, C_j) depends on the characteristic lifetime. It decreases if the difference in the characteristic lifetime is similar:

$$q_{dist,i,j} = -|a_i - a_j| \tag{4.26}$$

The quality qualifier value $q_{qual,i,j}$ for components (C_i, C_j) is depending on the difference between the expected lifetime of the two underlying Weibull distributions. As the MTTF (mean time to failure) is often used as a measure for component quality, this definition is a reasonable choice:

$$q_{qual,i,j} = \text{MTTF}_i - \text{MTTF}_j \tag{4.27}$$

The MTTF of a Weibull distribution is given by [137]:

$$\text{MTTF}_i = \frac{\gamma(1/\beta_i)\alpha_i}{\beta_i} \tag{4.28}$$

The wear out factor $q_{wo,i,j}$ is defined as the difference between the shape parameters. The higher the shape parameter, the steeper the rise of the failure rate. Therefore, the difference seems to be a sensible way to represent the property "wear out".

$$q_{wo,i,j} = \beta_i - \beta_j \tag{4.29}$$

All qualifiers were scaled to the interval $[-1, 1]$ and random Gaussian noise was induced.

4.5.2 Real Test Set

After the feasibility study and first investigation of the prediction methodology, it is necessary to validate it on a real test case. For this purpose, failure data of four different automatic transmissions was provided by ZF Friedrichshafen AG. The dependability data was derived from the calendared work-shop stops. The transmissions have been in the field for long. Data sets have been chronologically collected over a period of 8 years. Four different types of automatic transmission (T1-1, T1-2, T2-1, T2-2) were investigated, each of them with at least 200 workshop records containing the mileage until failure. The failure CDFs were fitted against the mileage in km (denoted t) by the Kaplan-Meier estimator [33], the number of censored units (units in the field) was estimated. Figure 4.8 shows the empirical failure distributions and the corresponding qualifier values. The derived functions have different ranges in both x-axis (mileage) and y-axis (failure probability) direction which raises the prediction difficulty. The estimated qualifier values for each of the 16 possible combinations of the four transmissions T1-1, T1-2, T2-1, and T2-2 are also shown in the figure. Transmissions T1-1 and T1-2

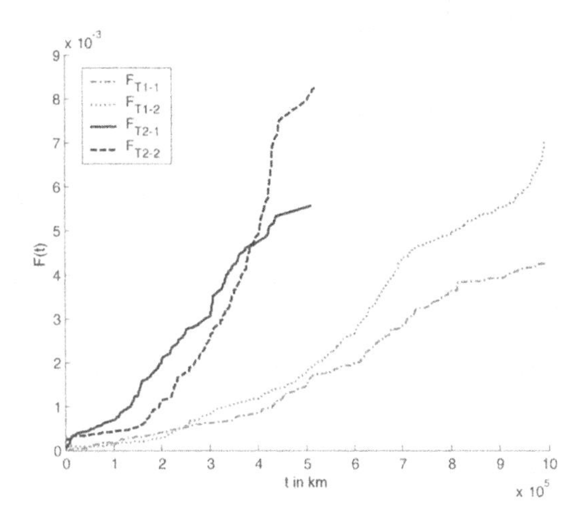

C_A	C_X	q_{dist}	q_{qual}	q_{wo}
T1-1	T1-1	1	0	0
	T1-2	0.9	0.3	0.2
	T2-1	0.3	0.8	0.1
	T2-2	0.4	0.9	0.5
T1-2	T1-1	0.9	-0.3	-0.2
	T1-2	1	0	0
	T2-1	0.4	0.6	0
	T2-2	0.3	0.8	0.4
T2-1	T1-1	0.3	-0.8	-0.1
	T1-2	0.4	-0.6	0
	T2-1	1	0	0
	T2-2	0.8	0.1	0.3
T2-2	T1-1	0.4	-0.9	0.4
	T1-2	0.3	-0.8	0.3
	T2-1	0.8	-0.1	0.8
	T2-2	1	0	1

Fig. 4.8. Failure distributions of four different automatic transmissions. Qualifier values for combinations of T1-1, T1-2, T2-1, T2-2.

are similar in their design, as are T2-1 and T2-2. The similarity is well founded by the fact that the transmissions are designed for similar torque. Types T1 and T2 show no strong similarities.

4.6 Results

In this section, it is investigated how both prediction methods perform on the test sets. The results are shown graphically in plots similar to figure 4.9.

The black line represents the function of the old component, which is used to predict the new function. It is revealed in the prediction process together with the qualifier vector. The blue line shows the best estimate $\check{F}_X$ on the function of the new component predicted by the NN / GP. The prediction bounds $\underline{F_X}$ and $\overline{F_X}$ (red), which are enveloping the best estimate give the estimated prediction uncertainty. If they are narrow, the NN / GP is confident to provide a correct estimate. In case of wide bounds, the trained method considers its predictions as highly uncertain. The green line, which was neither revealed in the training nor in the validation phase, shows the real function to be predicted. An ideal result would be that it is close to the best estimate, showing little uncertainty. On the other hand, a prediction with small uncertainty bounds that does not predict the correct function is worse than a function with larger prediction bounds.

4.6.1 Scalable Test Suite

The scalable test suite was used to test both prediction methods on a large data set. Including random noise in the functions and in the qualifiers, it serves as an almost realistic base for the prediction task. Two functions were to be predicted (highlighted

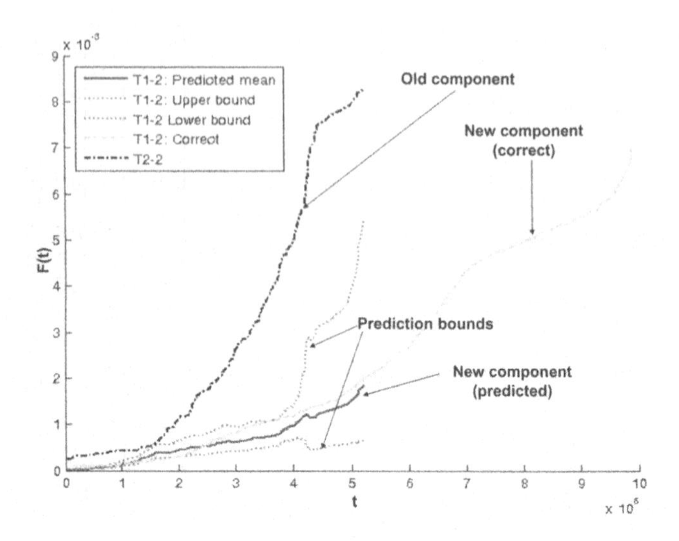

Fig. 4.9. Visualization of the prediction results. Black line: function of the old component, used to predict the new function. Blue line: predicted function of the new component. Red lines: prediction bounds (red) enveloping the prediction. Green line (not revealed in the prediction): real function to be predicted.

in figure 4.7): W5, which is one of the intermediate functions in the test set, and W10, which is a more extremal case. Functions to be predicted were not revealed in the training process. W5 and W10 were to be predicted using an extremely high failure rate (W2), a medium rate (W4 resp. W5) and a low failure rate (W10 resp. W9).

4.6.1.1 Neural Networks

The NN was evaluated on all six test cases. The results are plotted in figures 4.10 a)-f) on page 71. The predictions on W5 are acceptably accurate. The prediction of W5 via W2 (figure 4.10 a)) is accurate with little uncertainty and little error. W4 and W5 show a steep rise at the beginning (0 - 30,000), which the network is not able to reproduce (figure 4.10 b)). In the other regions, the estimation is accurate. The prediction using W10 as known component includes the real function entirely in the prediction bounds, but contains a high amount of uncertainty (figure 4.10 c)). The NN was not able to capture the exact shape of W5 but was able to "recognize" this behavior and predicted the uncertainty well.

Moving forward to the extremal cases, it turns out that the prediction of W10 using W2 was managed well (figure 4.10 d)) while predicting W10 from W5 was more difficult (figure 4.10 e)). The exact shape of W10 could not be predicted. However, by extending the width of the prediction bounds, more uncertainty is contained in the prediction. Thus, the user is warned that the best estimate may not be very accurate. The prediction of W10 using W9 is wrong, not including the real function in the bounds (figure 4.10 f)).

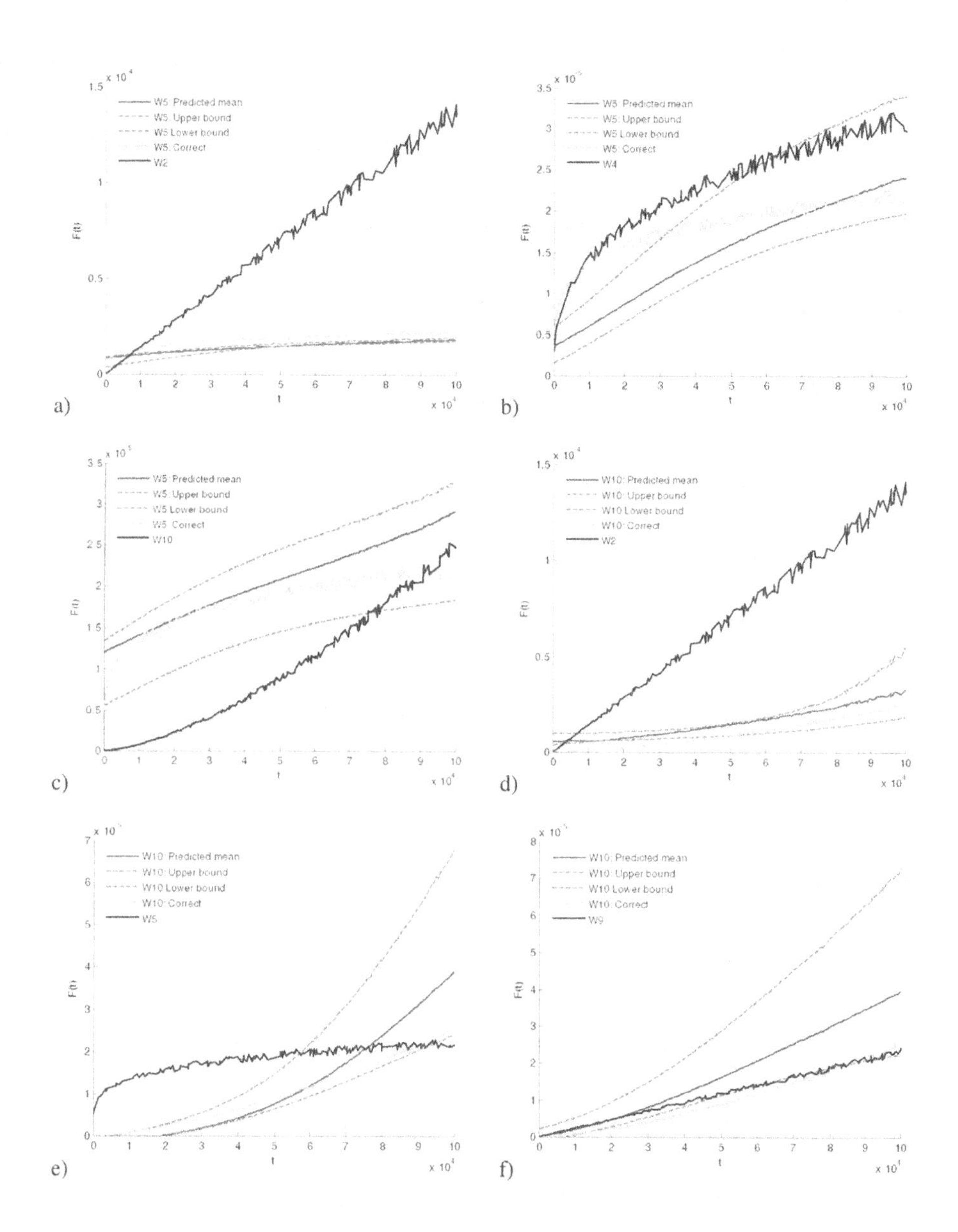

Fig. 4.10. Prediction results of the NN on the scalable test suite. a) Prediction of W5 using W2. b) Prediction of W5 using W4. c) Prediction of W5 using W10. d) Prediction of W10 using W2. e) Prediction of W10 using W5. f) Prediction of W10 using W9.

4.6.1.2 Gaussian Processes

The Gaussian processes as a concurrent approach to the neural network showed both positive and negative differences on the test cases. Results are depicted in figures 4.11 a)-f) on page 72. The prediction of W5 from W2 (figure 4.11 a)) is quite accurate. The

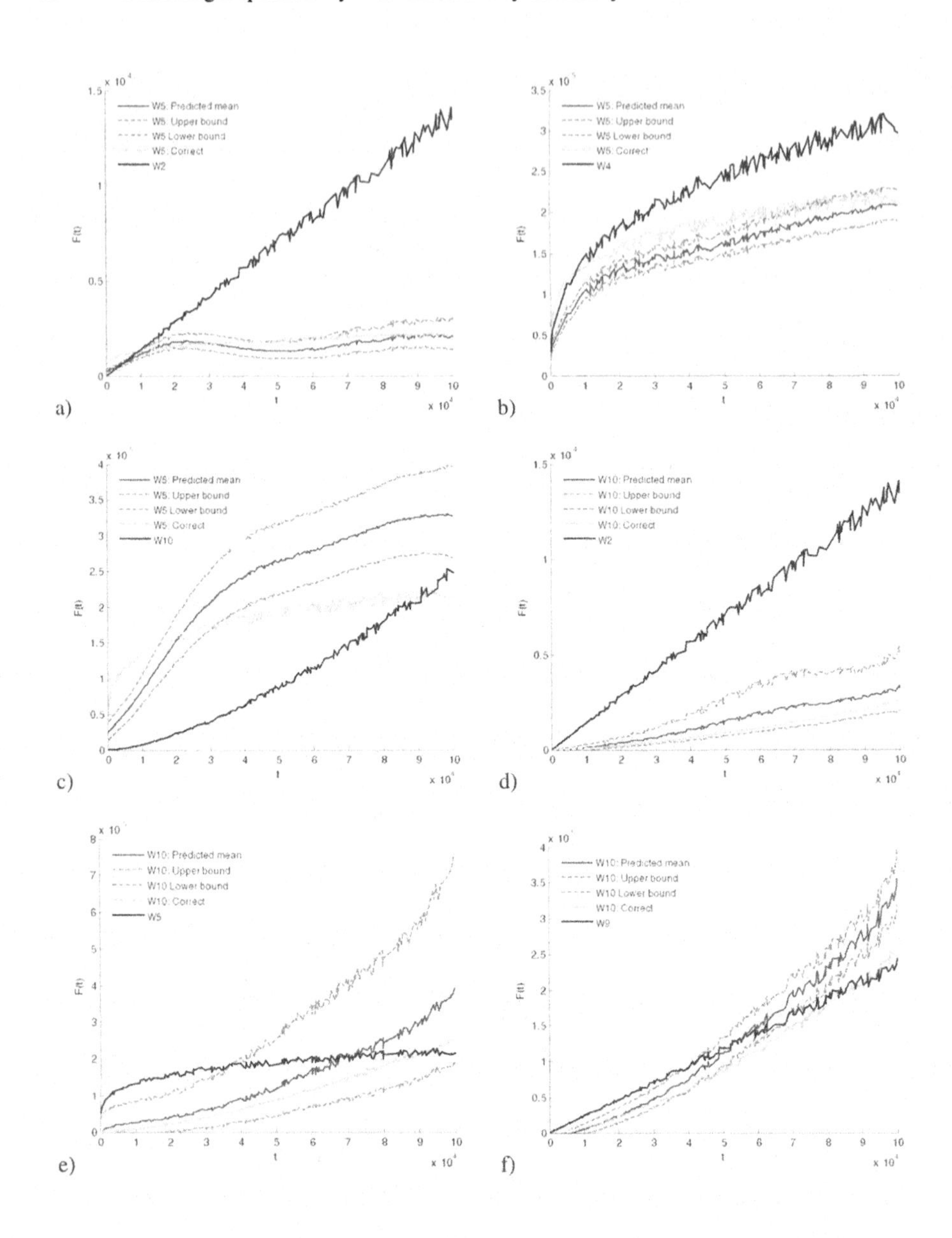

Fig. 4.11. Prediction results of the GP on the scalable test suite. a) Prediction of W5 using W2. b) Prediction of W5 using W4. c) Prediction of W5 using W10. d) Prediction of W10 using W2. e) Prediction of W10 using W5. f) Prediction of W10 using W9.

prediction of W5 from W4 (figure 4.11 b)) was not successful, the predicted function is too low. However, even if the Gaussian processes did not manage to predict the actual values, the predicted shape corresponds much better with the real function. The same situation is shown in figure 4.11 c), where W5 is predicted from W10. The results are worse than the ones obtained by neural networks, the function shape is closer to the correct one.

W10 was predicted well from W2 (figure 4.11 d)), including the real function completely in the bounds while keeping a medium amount of uncertainty. The prediction of W10 via W5 (figure 4.11 e)) is equally good, including a large amount of uncertainty but being more accurate than the corresponding neural network estimate. The prediction of W10 using W9 (figure 4.11 f)) is close to the real value, but the exact function was not reached. Nevertheless, in this example, the Gaussian processes also performed better than the neural network.

Subsuming the results for the artificial test set, both methods performed well in predicting unknown functions. The neural networks had difficulties in predicting W10, but managed to predict W5 well. On the contrary, the Gaussian processes gave appropriate predictions for W10, but were outperformed by the neural networks for the intermediate W5 function.

4.6.2 Real Test Set

Regarding the promising results on the artificial test set, the question which arises is how the prediction approaches perform on the real test set introduced in section 4.5.2. Two different scenarios were tested. The first one aims to test the accuracy of the method for predicting a function that is rather similar to one of the examples. The transmission T1-2, which has a failure CDF similar to T1-1 was chosen as an example. Both the NNs and the GPs were trained with a training set containing only T1-1, T2-1 and T2-2.

A second trial aims on testing the capability to predict the uncertainty bounds. Function T2-2 acted as the unknown function not revealed in the training process. In the training set, there is no function which is similar to T2-2. The estimated qualifiers in combinations which include C_X=T2-2 (figure 4.8) are different from all known training patterns. Thus, the NN / GP has no possibility of predicting the new function with high accuracy. In this case of a prediction task that is very different to all training cases, the NN / GP should "recognize" this difficulty and the bounds should be wide.

4.6.2.1 Neural Networks

From the results of the real tests (figures 4.12 a-f) on page 74), it can be seen that the neural network was able to predict the unknown functions quite well. Function T1-2 (figure 4.12 a)) was predicted with high accuracy from T1-1. Only in the first 200,000km region, the predicted bounds are too narrow. The predictions of T1-2 from T2-1 (figure 4.12 b)) were surprisingly good, too. With very narrow bounds, the results enclose around 75% of the real failure function. Only between 120,000km and 260,000km, a slight deviation is visible. Even if predicting T1-2 from T2-2 as shown in figure 4.12 c), the results are acceptable.

In the second test series, where T2-2 was to be predicted, it is investigated if the network can react with wider prediction bounds on unknown situations. The qualifiers indicate that the function to predict is not very similar. An adequate reaction would be

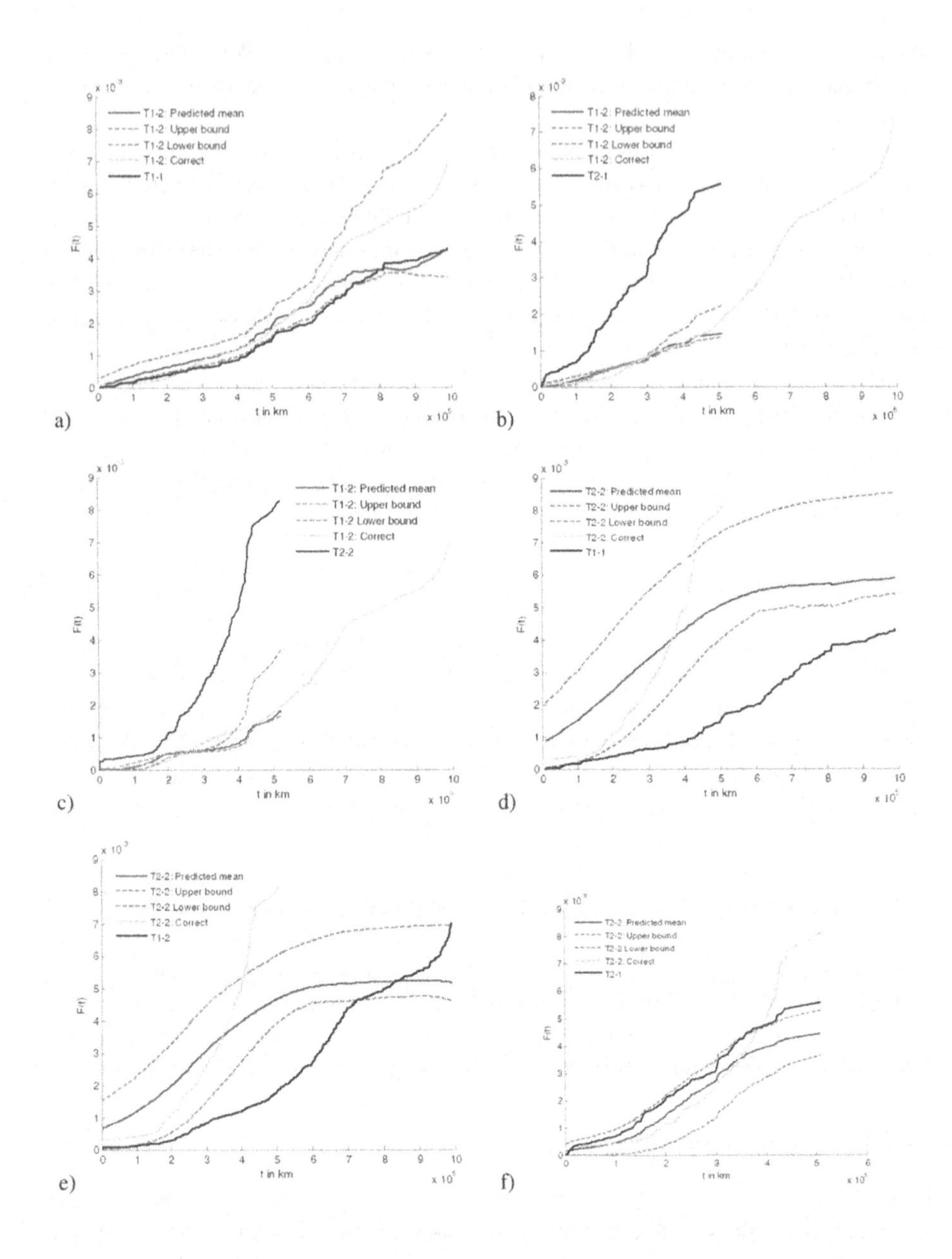

Fig. 4.12. Prediction results of the NN on the real test set. a) Prediction of T1-2 using T1-1 b) Prediction of T1-2 using T2-1. c) Prediction of T1-2 using T2-2. d) Prediction of T2-2 using T1-1. e) Prediction of T2-2 using T1-2. f) Prediction of T2-2 using T2-1.

to widen the prediction bounds. The most difficult case is to predict T1-1 from T2-2 (figure 4.12 d)). The bounds are far wider than in the first test series. The network is therefore not only able to provide correct predictions on the function, but also to predict

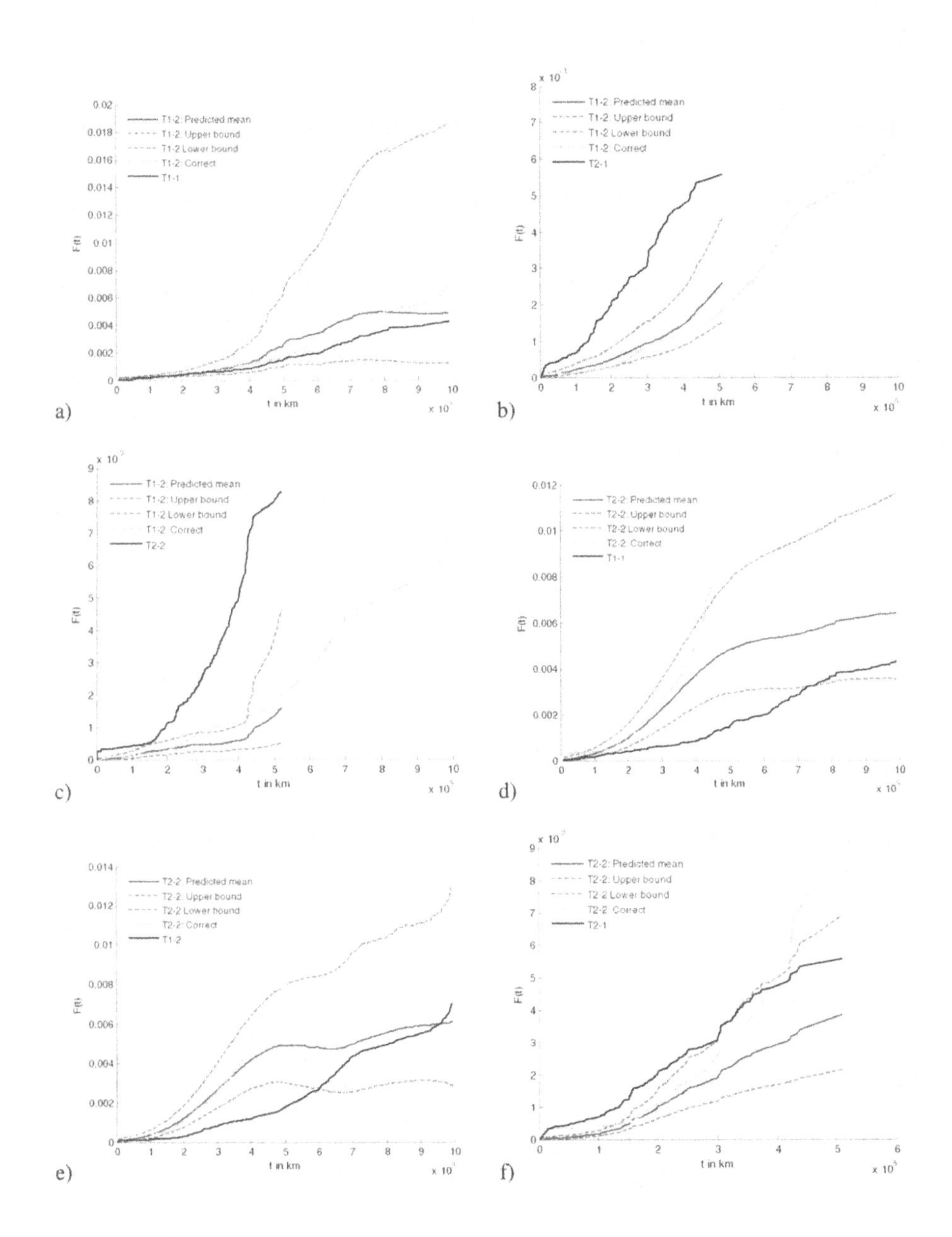

Fig. 4.13. Prediction results of the GP on the real test set. a) Prediction of T1-2 using T1-1 b) Prediction of T1-2 using T2-1. c) Prediction of T1-2 using T2-2. d) Prediction of T2-2 using T1-1. e) Prediction of T2-2 using T1-2. f) Prediction of T2-2 using T2-1.

the uncertainty around the best estimate. The second and third tests (T2-2 from T1-2, figure 4.12 e) resp. T2-2 from T2-1, figure 4.12 f)) show similar behavior. Therefore it can be summarized that for the real test set, the network was able to provide both a precise and robust estimation of the unknown component's failure properties.

4.6.2.2 Gaussian Processes

The Gaussian processes prediction (figure 4.13 on page 75) performed very well in predicting the T1-2 function. In figure 4.13 a) the prediction using T1-1 as known component can be seen. The best estimate is very close to the real function. The uncertainty bounds are quite wide but enclose the function T1-2 perfectly. Figures 4.13 b) and 4.13 c) show the prediction using T2-1 respective T2-2 as inputs. Both results are still very accurate. The best estimate is very close to the correct function of T2-2. The prediction bounds are narrow but enclose the correct function on a wide range. Especially for T2-2, which is strongly different from T1-2, this is remarkable. It shows that the GP approach has not simply learned to envelope F_A but has captured the qualifiers' meaning. Stating that T1-2 has a higher quality than T2-2 shifts the bounds to regions with a lower failure probability. It is necessary to mention that due to the input strategy (inputting time slices), a function value is predicted at each point of time of the original function. The domain of the predicted function is therefore identical to the domain of the input function and the predictions in figure 4.13 b) and c) only cover the domain of T2-1 respective T2-2.

The second test series attempt to predict the T2-2 function. The qualifier values are extreme, especially the distance qualifier. Therefore the GP approach is expected to react with wide bounds on the predicted function. Moreover, there was no similar training function which made the prediction even more difficult. Figure 4.13 d) shows the results for predicting T2-2 from T1-1. The GP is not able to predict the shape of T2-2. However it can be determined from the input data that no precise prediction can be made. The predicted uncertainty bounds are large and enclose T2-2 until 410,000 km. A similar result is obtained if using T1-2 (figure 4.13 e)). While maintaining a low amount of uncertainty, the prediction using T2-1 is more accurate (figure 4.13 f)), capturing the correct function completely in the uncertainty bounds. From the results can be seen that Gaussian processes are quite dependable in the prediction task and able to place the uncertainty bounds well.

4.7 Conclusion

This chapter has presented a novel way of dependability prediction. In-service data and expert estimates were combined to a single prediction method. Instead of predefining a prediction algorithm, two different methods from computational intelligence were used to regress the prediction function from training data. Both neural networks and Gaussian processes were able to solve the prediction task. The prediction method preserves elicitation uncertainty and thus allows robust dependability modeling. Due to the high availability of expert knowledge during the design process, the methodology is especially suitable in EDS. Practical applicability was proved by successfully predicting artificial and real failure functions (failure rates, CDFs). In further work, it has to be investigated if other machine-learning methods are able to raise prediction performance. But perhaps most interesting will be the application and evaluation of the method on a larger data set or data base.

5 Design Space Specification of Dependability Optimization Problems Using Feature Models

Predicting EDS dependability is not an end in itself. One of the main reasons for quantifying the dependability of a system is the need for some comparable values to find an appropriate system design. Dependability-driven system optimization has been for long time a topic of considerable interest. Various objectives such as the dependability constituents reliability and safety, cost, weight and prediction uncertainty were chosen to rate the systems. Many search heuristics such as evolutionary algorithms [30], tabu search [96] and physical programming [161] have been applied to screen the design space for interesting systems regarding objectives as dependability and costs (a more detailed overview is given in section 5.1).

However, the choice of the right optimization strategy is only one aspect in the system optimization process. Other aspects include the definition of one or several objectives of interest, their nature (deterministic, probabilistic) and their uncertainty. Perhaps even more important is the specification of a feasible design space fitting the requirements of the decision-maker. From the infinite set of all possible systems, dependability engineers need to select the set of potentially interesting ones, the design space.

Defining the design space for a dependability optimization problem can be considered as selecting a subset D from the space of all possible system models S (figure 5.1). The design space contains all systems that can be screened by the optimization algorithm for optimal system designs. D is ideally the set including all systems which are of potential interest to the decision-maker, but excluding all other systems. Exceedingly large sets slow down the optimization process and may end up in results that do not meet these interests. Sets that are too small may exclude potentially optimal solutions beforehand. However, the method to specify D must have a very low complexity in itself. It is not feasible to select each system of interest manually, considering that D may contain millions of system variants. Only if design spaces with high complexity and complex constraints can be defined with little effort, the automatic screening and optimization methods can be useful. A method that allows the rapid specification of D without imposing restrictions on the shape of D may be a great help to the decision maker. However, exactly this question has been treated only very little in literature and will be the subject of this chapter. There is no single optimal solution, as high flexibility

P. Limbourg: Dependability Modelling under Uncertainty, SCI 148, pp. 77–88, 2008.
springerlink.com

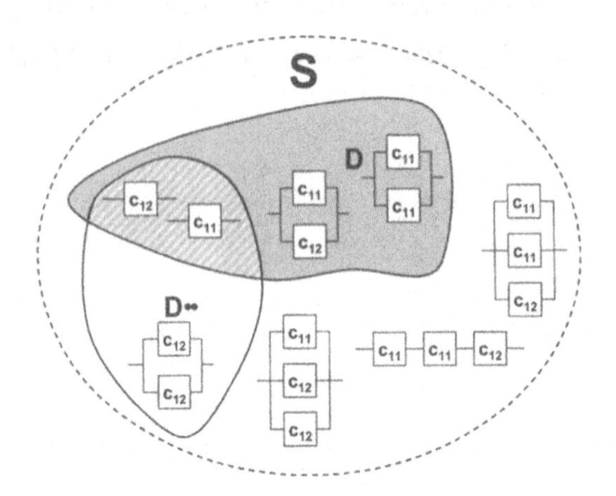

Fig. 5.1. Schematic view of the system space (S) and two design spaces ($D \subseteq S, D' \subseteq S$). Each design space contains a set of possible system designs.

and low complexity are conflicting goals. Feature models, the modeling approach used in this chapter, are a good trade-off, being a powerful, flexible modeling tool that is at the same time intuitively in its use.

A common and intuitive approach to define the design space is the redundancy allocation problem (RAP). It has been covered e. g. by [30, 96, 161, 29, 118, 101] and [28] (a more detailed presentation of these references will be given in sections 5.1 and 6.1.2.1). However, its practical applicability is limited to a small set of optimization problems. The degrees of freedom in the modeling process are rather small. It may be difficult to specify complex design spaces without including solutions that are a priori unwanted. Realistic design problems are complex mixtures which may include the decision between different components, nested subsystems with further variability, redundancy and other fault tolerance patterns in a single optimization problem.

The presented approach for optimizing system structures is a generalization of RAP optimization. In figure 5.2, the components and their interactions are shown. The main tool for the design space specification is feature modeling [36], a technique stemming from software engineering [37]. In this work, a novel use for feature models not for the representation of existing systems but for the fast specification of large sets of system alternatives in optimization scenarios is proposed. Combined with a vector-based enumeration scheme (realization generator), feature models can represent the decision space of a combinatorial optimization problem. This space is searched by an optimization algorithm testing single realizations of the model. Realizations are converted to fault trees and reliability block diagrams (RBDs) by the RBD / fault tree generator and evaluated according to several objectives. With this new approach, it is possible to formulate dependability optimization problems with large complexity in an efficient way.

The remainder of this chapter covers the components in figure 5.2 that deal with the specification and realization of the design space (dashed box). Section 5.1 describes a popular form of system dependability optimization, the redundancy allocation problem.

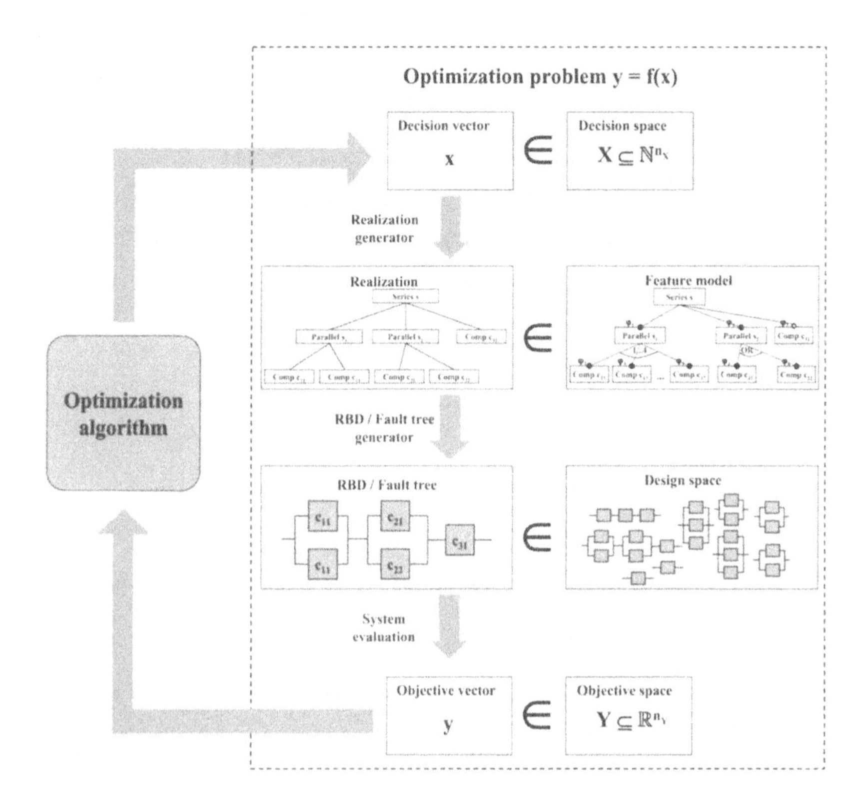

Fig. 5.2. Optimization through feature modeling - overview

In section 5.2, the basic concepts of feature modeling are introduced. Section 5.3 illustrates the different feature sets and proposes some additional feature set types useful for dependability optimization. Sections 5.4 and 5.5 apply feature models to the dependability optimization case. In section 5.4, it is shown how optimization problems can be defined by a feature model combined with an enumeration scheme. Section 5.5 describes a way of converting a configuration of the model to reliability block diagrams and fault trees and shows, how arbitrary redundancy allocation problems can be described with feature models.

5.1 The Redundancy Allocation Problem

The better part of work regarding system dependability optimization tackles the right placing of component redundancies. The redundancy allocation problem (RAP) is a way to set up the design space according to this question. A system with n subsystems $s_1,...,s_n$ (figure 5.3) which are commonly put in series is the backbone of the formulation. Each subsystem s_i consists of m_i parallel placeholders $s_{i1},...,s_{im_i}$x which contain exactly one component from a set of possible component types $C_i = \{\emptyset, c_{i1},...,c_{ik_i}\}$

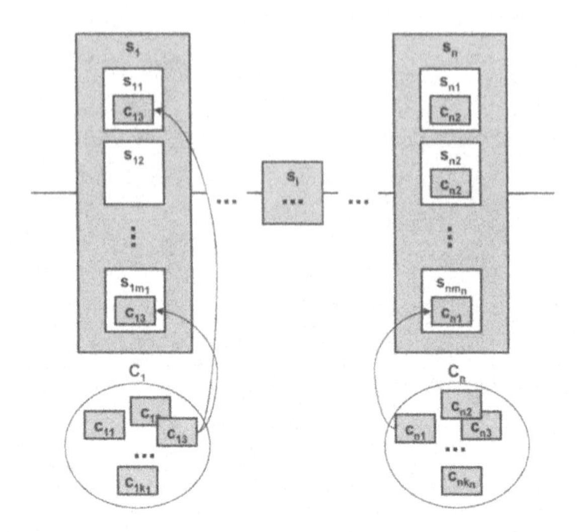

Fig. 5.3. The redundancy allocation problem. Each subsystem s_i can contain $1...m_i$ parallel components.

including $\emptyset$ as the alternative "no component". Several placeholders may contain the same component type.

Finding the optimal system in a RAP design space has been proved to be NP-hard [22] and since then there has been a considerable effort to solve this problems with various heuristics and different objective functions (section 6.1.2.1).

The RAP is very straightforward, easy to implement and to visualize. However, there are limitations: A placeholder can only be filled with components. It is difficult to model "recursive variability", the choice between subsystems with further internal variability. The same holds for the definition of constraints such as "either component c_{11} with double redundancy or component c_{12}", which are needed for realistic scenarios. Dependencies in the selection of different building parts are not uncommon but cannot be modeled in a RAP. It may not be possible to put components of different vendors in parallel redundancy. Or there may be the choice between a third-party component and a subsystem with further variability.

[20] solve this problem by modeling a RAP on the high level, and give a list of all possible configurations per block. This approach is viable for a "flat" problem with little or no variability in the subsystem and subsubsystem level. If the problem contains complex subsystems containing further variability, size and complexity of the model become very large.

In contrary approaches, the design space is extended to a wide range of system structures. The advantage of these approaches is that optimal structures can be found, which are not covered by a RAP coding. On the other hand, the design space may become incredibly large, wasting search time on a priori uninteresting regions.

In [100], a linear coding for arbitrary series-parallel system is proposed. It is very general in the sense that all possible series-parallel systems with a specified set of components are included, regardless of the underlying structure function. The design space is very large, but difficult to constrain to the interesting regions. In [53] and [52], optimization algorithms are operating directly on a system structure graph. The solutions are not represented as vector optimization problems, but by a graph representation. The resulting design space is almost constraint-free, allowing for a broad spectrum of systems. However, it is difficult to impose constraints and restrict the search on the promising regions of the design space. Because of the graphical representation, vector-based optimization algorithms cannot be applied. Genetic algorithms with special operators capable of modifying graph structures are used.

A desirable design space specification combines the advantages of these different approaches, providing a representation of the design space that can be flexibly tailored to the requirements of the user. All interesting systems should be reachable, but as little systems which are a priori uninteresting should be included, thus speeding up the search process. The proposed approach can be seen as a generalization of the RAP representation towards graph-based / tree-based representation, providing both a way for the rapid specification of design constraints and a flexible design space. It is not a way to code the whole design space (or the largest part) into a representation that can be used by a genetic algorithm. On the contrary, feature models allow to constrain the design space with a large flexibility to the systems that the user considers as important (it is shown in section 5.4, that RAP design spaces can be represented by special types of feature models), thus distinguishing this approach from [100] and [53].

5.2 Feature Models

Having outlined the restrictions imposed by the RAP formulation, it is necessary to focus on a more general approach to system modeling. While there is no design space specification method in dependability engineering which is both simple and powerful, other scientific areas provide a ready-made solution. In the field of software engineering, feature modeling [38] is used for managing families of configurable software. With the increasing need for managing multiple configurations of software tailored to the customer requirements, methods for the systematic definition of configuration sets or system families gained importance.

Feature modeling is a graphical method for set specification. Its purpose is to identify commonalities among the systems of interest while managing the differences between them in a systematic way. By selecting feature configurations, new systems may be realized from the specified system family. In [38], feature modeling is defined as:

Definition 5.1 (Feature modeling). *Feature modeling is the activity of modeling the common and the variable properties of concepts and their interdependencies and organizing them into a coherent model referred to as a feature model [38].*

As can be seen from this definition, feature modeling aims on modeling sets of concepts using their commonalities and differences. The result of this activity is a model

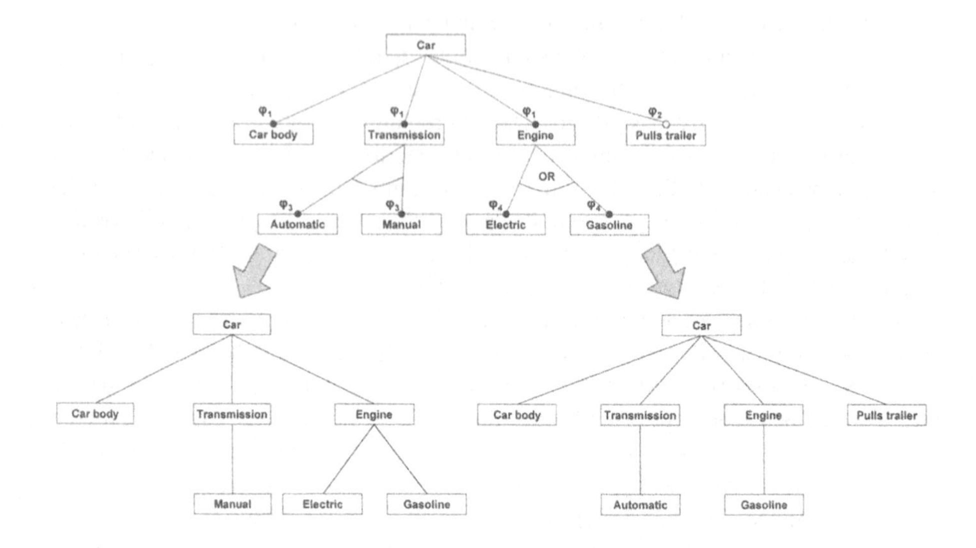

Fig. 5.4. Feature model of a car and two possible realizations (adapted from [38]). The feature model describes a set of design alternatives for a car. Both realizations are elements of this set.

describing all concepts in a systematic and compressed form. Concepts are defined as *any elements and structures in the domain of interest* [38]. Considering the task, the domain of interest is the system space S, while the concepts are the systems $s \in D$ represented as fault trees or reliability block diagrams. The feature model, describing the variable and common properties of a set of concepts is nothing else than a description of the design space $D \subseteq S$. Feature models are organized in a hierarchic, treelike form. The root of a feature model is referred to as the *concept node*. All other nodes are *feature nodes*. Nodes may contain information such as the name or other properties of the feature. Parent and child nodes are connected with directed edges. The edges are grouped to different types of feature sets. These types define if the child features are realized given that the parent feature is realized. A realization or instance description of a feature model is a tree containing the concept node and some or all feature nodes. By traversing the feature tree and configuring each feature set (choosing one of the feature realizations), a valid realization is reached.

Figure 5.4 shows a feature model of a simple set of cars. The concept node *car* has four feature nodes: *Car body*, *Transmission*, *Engine* and *Pulls trailer*. *Transmission* and *Engine* each have two feature nodes grouped to a feature set: *Automatic* / *Manual* and *Electric* / *Gasoline*. The feature model captures several common-sense restrictions for realizations. There is no modern car without a car body, a transmission and an engine. If *Car* is realized, these features are realized, too. However *Pulls trailer* is an optional feature of a car. Transmissions are either *Automatic* or *Manual*, but not both. These features are exclusive. The engine may be *Electric*, *Gasoline* or a hybrid (both *Electric* and *Gasoline*).

Such statements may be represented by a feature model using the different feature set types explained in the next section. The car feature model represents twelve different realizations of the concept car. Figure 5.4 moreover shows two possible realizations that emerge if all feature nodes are configured.

5.3 Basic Feature Set Types

There are several feature set types which define if and how often child feature nodes $v_1, \ldots, v_{|ch|}$ (with $|ch|$ being the number of child feature nodes) are realized, given that the parent feature node v_p is realized. If the parent is not realized, the child feature nodes are never realized. In [36], several basic feature set types are defined. The notations φ

Table 5.1. Feature types, symbol and number of different configurations

Type	Symbol	Description	Number of possible realizations $\lvert\varphi\rvert$
Mandatory	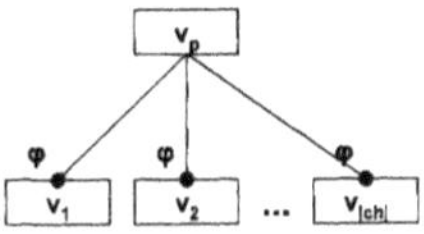	Select all	$\lvert\varphi\rvert = 1$
Dimension	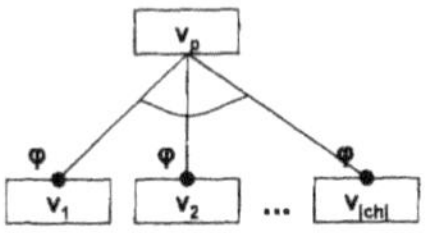	Select one	$\lvert\varphi\rvert = \lvert ch\rvert$
Dimension with optional features	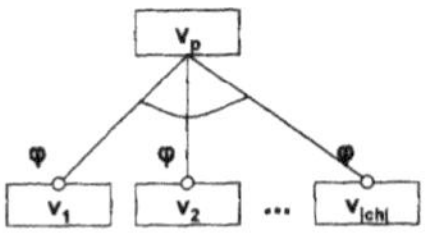	Select 0 or 1	$\lvert\varphi\rvert = \lvert ch\rvert + 1$
Optional	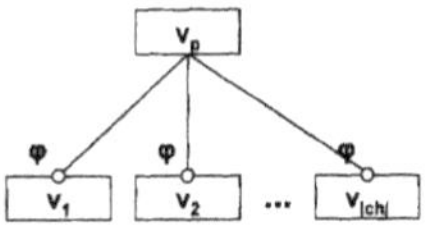	$0 \ldots \lvert ch\rvert$ arbitrary children may be selected	$\lvert\varphi\rvert = 2^{\lvert ch\rvert}$
OR	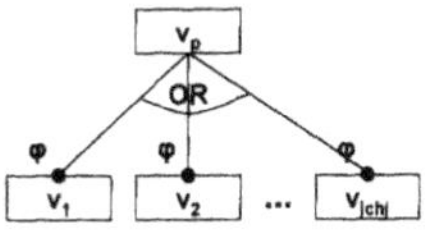 OR	$1 \ldots \lvert ch\rvert$ arbitrary children may be selected	$\lvert\varphi\rvert = 2^{\lvert ch\rvert} - 1$
Select-m	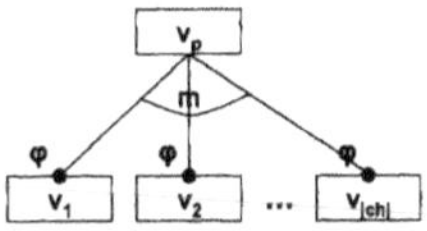 m	Select m features. Multiple selection allowed	$\lvert\varphi\rvert = \binom{\lvert ch\rvert + m - 1}{m}$
Select-1:m	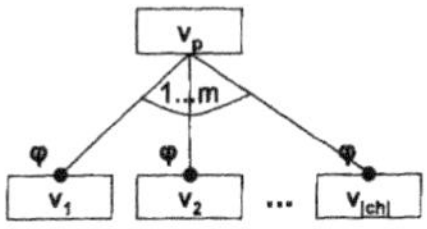 1..m	Select 1:m features. Multiple selection allowed	$\lvert\varphi\rvert = \binom{\lvert ch\rvert + m}{m} - 1$

for a feature set and $|\varphi|$ for its number of different configurations will be used in the following.

Mandatory feature set. All nodes in a mandatory feature set are included in the system realization if v_p is included. This feature set does not represent a variability in itself but helps to structure the system.

Dimension. A dimension denotes a set of mutual exclusive alternatives of which one needs to be chosen. If v_p is included, exactly one child has to be selected.

Dimension with optional features. This feature set represents an optional inclusion of mutual exclusive features. If v_p is included, none or one child can be included.

Optional feature set. Each child may be included in the realization, if v_p is included. Thus, zero, one or any combination of optional feature nodes can be included.

OR feature set. This feature is similar to the optional feature set. If v_p is included, any nonempty combination of children will be included.

In this work, two additional feature types are proposed, which are useful for dependability optimization. They support the multiple selection of the same feature and allow to specify redundancy problems in an elegant way.

Select-m feature set. If v_p is selected, then any combination of exactly m children may be included. In contrast to the previous feature set types, it is possible to select a child node several times.

Select-1:m feature set. A feature set similar to select-m. If the parent node is selected, any combination of 1 to *m* children may be included (multiple selection allowed).

An overview on the graphical representation of these feature sets and their number of possible realizations is given in table 5.1. Returning to the car example, *Pulls trailer* is an optional feature set. *Electric* and *Gasoline* belong to an OR feature set, *Automatic* and *Manual* to a Dimension.

5.4 Feature Models Defining Optimization Problems

After constraining D with a feature model, it is necessary to find a vector-based description of this space. This description, referred to as the representation, is a surjective mapping from the decision space X to the design space. All possible designs in D should be described by at least one vector in X (surjectivity). The chosen representation of a feature model maps each feature set with more than one realization to one design variable. Let $\Phi^+ := \{\varphi : |\varphi| > 1\}$ be the set of all features with more than one realization. The features sets $\varphi_1^+, ..., \varphi_{|\Phi^+|}^+ \in \Phi^+$ are numbered according to their appearance in a preorder traversal of the feature model. For each feature set φ_i^+, a corresponding variable $x_i \in \{1, ..., |\varphi_i^+|\}$ is included in the decision vector. Each value of x_i represents one configuration of φ_i^+. The decision vector x with dimension $|\Phi^+|$ is composed of $x_1, ..., x_{|\Phi^+|}$:

$$x \in \mathbb{N}^{|\Phi^+|} \tag{5.1}$$

$$x := \begin{pmatrix} x_1 \in \{1, \ldots, |\varphi_1^+|\} \\ \vdots \\ x_{|\Phi^+|} \in \{1, \ldots, |\varphi_{|\Phi^+|}^+|\} \end{pmatrix} \tag{5.2}$$

Each decision vector x defines a feasible realization for all features sets and therefore a feasible realization of the feature model. Furthermore, each combination of feature set realizations and therefore each realization of the feature model can be described by at least one decision vector.

The representation should support a fast progress of the optimization routine. This property can't be proved and strongly depends on the heuristic applied. For evolutionary algorithms, there are at least two rules of thumb.

First, the representation should support the generation and multiplication of "building blocks" [63] with high fitness to achieve effective crossover operations. The performance of evolutionary algorithms heavily depends on the ease of finding such building blocks (neighbored elements in the decision vector) which form patterns that lead to good solutions. These building blocks can be entirely swapped by the crossover operator and thus multiplied in other solutions.

Second, higher-order dependencies between design variables and objectives should be low to increase the quality of mutations. Mutations randomly change certain variables of the decision vector, leaving others almost untouched. If the dependency pattern is easier, then the performance is higher.

This representation supports the formation of building blocks by a genetic algorithm. Due to the preorder sorting of $x_1, \ldots, x_{|\Phi^+|}$, components / subsystems belonging to the same subsystem occupy neighbored positions in x. Good subsystems therefore correspond to successful linear building blocks and have a good chance to be exchanged entirely by the crossover operator. If the design variables are largely independent, pointwise mutation operators increase their efficiency. This property is at least partly fulfilled. Each feature set is represented by exactly one design variable and feature sets are independent from another.

The mapping is surjective, but not bijective: several vectors in X can describe an element in D. If a feature is not realized, values for unreached child feature sets may vary arbitrary without changing the realization. However, this is no drawback and may even be an advantage for evolutionary algorithms. Redundancy in input coding is a biologically-inspired technique that is often applied to achieve better optimization results [132].

5.5 Generating Reliability Block Diagrams and Fault Trees from Realizations

If feature models are used to define reliability block diagram or fault tree families, it is necessary to build a generator that maps realizations of a feature model into this system models. Two attributes (name and type) are attached to each node. In case of a conversion to RBDs it is distinguished between the types *Series* representing a series subsystem, *Parallel* (parallel subsystem), *k-out-of-n* (k-out-of-n-good subsystem, hot standby redundancy) and *Comp* representing a component. Similarly, if the feature model represents a fault tree family, the types *AND*, *OR*, *k-out-of-n* and *Comp* exist. Other elements such as bridge structures and more complex subsystem types can be introduced without modifying the parsing algorithm. Nodes are named according to the component / subsystem they represent. *Comp* nodes must be leaves, while all others must be inner nodes. Realizations according to this scheme can be converted into a reliability block diagram / fault tree with a recursive preorder traversal (listing 5.1, page 87).

The time complexity of this algorithm can be deduced from a depth-first search (DFS) traversal. A feature model configuration v is per definition a tree [38]. The code described in listing 5.1 shows a DFS traversal in lines 12-15 with a recursive call for each adjacent child node $V[1,...,|ch|]$ in line 14. In [32] it is shown, that the overall number of recursive calls can be bounded by $O(n)$ for a tree with n nodes (one for each tree node).

Lines 16-25 are executed for each call of RBD-Fault tree-Generator. There is no recursion included. In the switch-case block, the system description string s is formed. Four different cases have to be investigated:

Component. The component is looked up in a sorted data structure with all components included. A project library of m components would therefore cause a complexity of $O(log(m))$, assuming that the components are stored in a sorted binary tree. Depending the component to the system description string is possible in $O(1)$ if e. g. using a stack.

Series system. The subsystem description strings are concatenated and attached to the system description string. The number of components of all subsystems is limited by the nodes in the feature model n. Concatenation is therefore possible in $O(n)$.

Parallel system. Identical to "Series system", $O(n)$.

k-out-of-n-good-system. Identical to "Series system", $O(n)$.

This leads to a complexity of lines 16-25 of $O(\max(n, log(m)))$ per call. The overall complexity for n calls is therefore bounded by $O(n \cdot \max(n, log(m)))$ for n nodes and a data base of m components.

The transformation of a feature realization to the corresponding RBD / fault tree can also be seen in the overview (figure 5.2). The RBD / fault tree corresponding to a decision vector x will be denoted as $s(x)$.

The RAP design space can be represented as a feature model in an elegant way (figure 5.5). Defining the root node as the series feature, the n systems $s_1, ..., s_n$ are included as mandatory features with type parallel. Each parallel feature s_i has child features for each component $c_{i1}, ..., c_{ik_i}$. All component features of subsystem s_i are joined by a select-1:m_i feature. This ensures that a maximum number of m_i components are included in parallel.

But feature models are not just another way to formulate the RAP. The advantage of feature models is that problems with higher complexity are tractable, too. Subsystems may be at the same level as components, the choice of the components may be restricted to certain combinations and the design space is not limited to a fixed amount of series subsystems. In section 6.5, a complex design problem illustrating these vantages is shown.

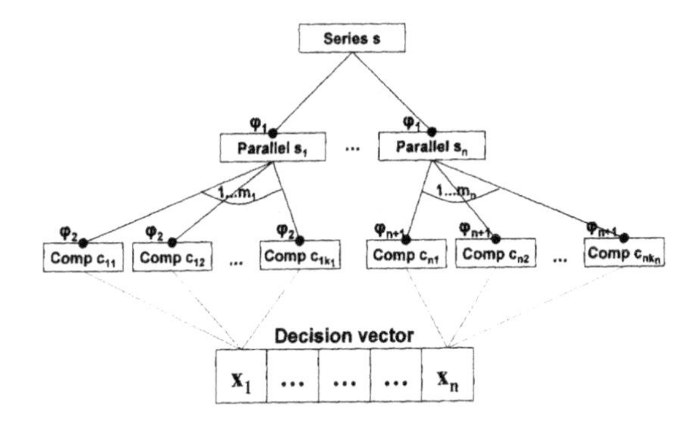

Fig. 5.5. The RAP as a feature model with corresponding decision vector

Listing 5.1. RBD / Fault tree generator algorithm

```
1  ALGORITHM RBD-Fault tree-Generator (v)
2  Description
3     Generates a RBD / Fault tree from a feature model configuration
4  Definitions & declarations
5     Receives: v (Sub-)tree of a feature model configuration
6     Returns: s
7     Set of subtrees: V
8     Integer: i
9     Set of systems: S
10    System: s
11 Core
12    V[1,...,|ch|] = all direct child nodes of v
13    FOR i = 1 TO |ch| STEP 1
14      S[i] = RBD-Fault tree-Generator(V[i])
15    END FOR
16    SWITCH v.type
17      CASE Series / AND
18        s = Series system of S[1,...,|ch|]
19      CASE Parallel / OR
20        s = Parallel system of S[1,...,|ch|]
21      CASE k-out-of-n-good
22        s = k-out-of-n-good system of S[1,...,|ch|]
23      CASE Comp
24        s = Component with name v.name
25    END SWITCH
26    RETURN s
27 END ALGORITHM
```

5.6 Conclusion

In this chapter, a novel way to define and refine the design space of system dependability optimization problems is presented. Feature modeling has been applied to allow a very flexible formulation of the optimization problem. With feature models, the design space can be formulated and tailored rapidly to the user's needs. This has been shown by formulating the redundancy allocation problem (RAP) and a more complex problem via feature models. It is shown, how feature models support vector-based enumeration of the design space, thus making it accessible as a decision space to arbitrary combinatorial optimization algorithms. With a simple, recursive algorithm, it is possible to convert realizations of the proposed feature models to block diagrams or fault trees.

This contribution was the first application of feature models on optimization problems in dependability engineering. If this approach will be extended in further research, possible starting points may be the investigation of other representations of the design space such as cardinality-based feature models [39].

6 Evolutionary Multi-objective Optimization of Imprecise Probabilistic Models

Multi-objective evolutionary algorithms (MOEAs) have become increasingly popular in the recent past to discover optimal or near optimal solutions for design problems with a high complexity and several objectives [26]. Hence, MOEAs seem to be an interesting option for dependability optimization in an EDS. However most MOEAs are designed for deterministic problems, requiring that an evaluation of the objectives for a solution results in a deterministic objective vector without uncertainty.

However, as discussed in chapter 3, a lot of reasonable effort may be put into the formulation of an uncertainty-preserving model to hedge the uncertainty. Especially in system dependability prediction, uncertainty modeling is a very important aspect in the overall prediction process. Therefore, if included into an optimization loop, the results obtained from the system model remain uncertain. Very little research has already been devoted to the adaption of MOEAs to uncertain objectives. In this chapter, all necessary means to construct a MOEA for uncertain objective functions are developed. The main focus is to investigate new ways for estimating density and obtaining the nondominated set that are transferable to a wide range of other popular MOEAs. Equipped with these uncertainty handling strategies, the novel MOEA may then be used in the context of system dependability optimization.

The chapter is structured as follows: in section 6.1, an introduction to Pareto-based multi-objective optimization for deterministic and uncertain objective functions is given. Section 6.2 introduces MOEAs in more detail and a base algorithm for optimization under uncertainty is developed. In section 6.3 and 6.4, the new contributions to adapt a MOEA to multiple uncertain objectives are presented. A dominance criterion and a new niching operator are proposed. Section 6.5 illustrates the whole methodology on two examples, the redundancy allocation problem (RAP) and a more complex optimization task. Section 6.6 gives a chapter summary and some further research directions.

6.1 Pareto-Based Multi-objective Optimization

This section introduces the concept of Pareto-based multi-objective optimization. Starting with deterministic multi-objective functions, concepts such as Pareto optimality

P. Limbourg: Dependability Modelling under Uncertainty, SCI 148, pp. 89–106, 2008.
springerlink.com

and dominance are introduced. Then it is shown how imprecise multi-objective functions can be described and Pareto optimality can be transferred and implemented under imprecision.

6.1.1 Deterministic Multi-objective Functions

Deterministic multi-objective optimization problem formulations are mappings from the decision space X to an objective space $Y \subseteq \mathbb{R}^{n_Y}$ with dimension n_Y where the goal is to find a vector x^* or set of vectors X^* that maximize the objective function f:

$$f : X \rightarrow Y \subseteq \mathbb{R}^{n_Y} \tag{6.1}$$

$$f(x) = \begin{pmatrix} f_1(x) \\ \vdots \\ f_{n_Y}(x) \end{pmatrix} = y = \begin{pmatrix} y_1 \\ \vdots \\ y_{n_Y} \end{pmatrix} \tag{6.2}$$

If $n_Y = 1$, then f is a single-objective optimization problem. If $n_Y > 1$, then the concept of "optimal" / "maximal" is different, because it includes a (partial) order over vectors.

The most desirable case that one solution maximizes all $f_1, \ldots, f_n$ simultaneously is not most likely attained. More often a set of solutions that are optimal according to a specific trade-off occur. The meaning of "optimality" in case of several objectives is therefore different. For deterministic multi-objective functions, a common optimality criterion is the Pareto dominance $\succ_P$ [126]:

$$x \succ_P x' :\Leftrightarrow \begin{cases} \forall i = 1, \ldots, n_Y : f_i(x) \geq f_i(x') \\ \exists i = 1, \ldots, n_Y : f_i(x) > f_i(x') \end{cases} \tag{6.3}$$

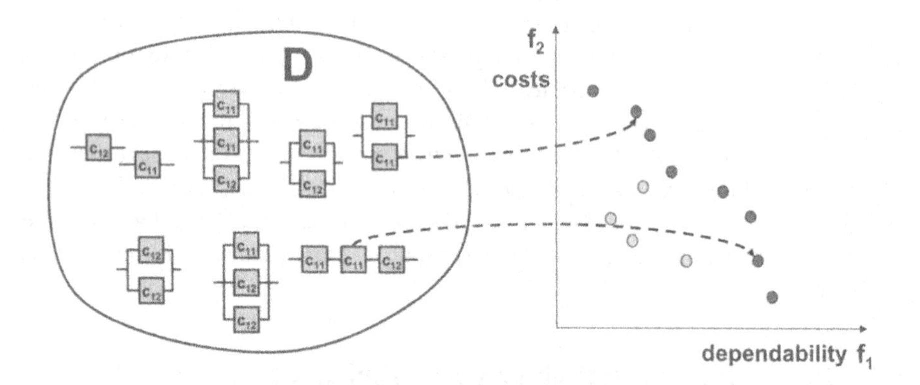

Fig. 6.1. An exemplary set of solutions in the objective space. Dark blue solutions are nondominated. Each solution is optimal for a certain trade-off between costs and dependability measure. Light blue solutions are dominated (nonoptimal for all trade-offs).

The verbal representation of $\succ_P$ is that solution x dominates x' if it is better in one objective without decreasing any other. A solution x^* is Pareto optimal, if no other solution dominates x^*:

Definition 6.1 (Pareto optimality). x^* *is Pareto optimal if:*

$$\nexists x' \in X : x' \succ_P x^* \tag{6.4}$$

Most likely, this property holds not only for one, but for a set of solutions $X^* \subseteq X$, the Pareto optimal set (or informally, the set of all optimal trade-offs). In figure 6.1, a set of solutions is illustrated. Each decision vector in the decision space (representing a system design) corresponds to a solution vector in the objective space. The dark solutions form a nondominated set. None of the solutions is better in all objectives than another.

6.1.2 Imprecise Multi-objective Functions

Clearly the above relationship is suitable for exact values, but what if the results of f entail imprecise values, represented by probability distributions or BPAs? If system dependability is to be optimized in an EDS, the objective functions are of an imprecise nature and in this case modeled as BPAs. Normal Pareto dominance concepts cease to work and extensions to handle such functions have to be introduced. In chapter 3, it was shown how for a given system, the system function ϕ produced uncertain results. If a dependability prediction in the Dempster-Shafer representation is included in a multi-objective optimization problem, objectives become imprecise. An imprecise-probability multi-objective function $\widetilde{f}$ can be defined as a function with an imprecise probabilistic output for a deterministic decision vector:

$$\widetilde{f}(x) = \begin{pmatrix} \widetilde{f}_1(x) \\ \vdots \\ \widetilde{f}_{n_Y}(x) \end{pmatrix} = \widetilde{Y} = \begin{pmatrix} \widetilde{Y}_1 \\ \vdots \\ \widetilde{Y}_n \end{pmatrix} \tag{6.5}$$

In the deterministic case, $f(x) = y$ is a deterministic value. If compared to another objective vector y', the expression $y = y'$ is either true or false. In the probabilistic case, $\widehat{f}(x) = \widehat{Y}$ is a probability distribution, and $P(\widehat{Y} = y') \in [0,1]$. In the imprecise case, this probability is bounded, $\widetilde{f}(x) = \widetilde{Y}$ is a multidimensional BPA, and the equality to a vector y' can only be given by $Bel(\widetilde{Y} = y') \in [0,1], Pl(\widetilde{Y} = y') \in [0,1]$.

To illustrate the different objective types, figure 6.2 shows a probabilistic (left) and an imprecise probabilistic (right) multi-objective solution. In the probabilistic case, the solution is not a point, but a joint distribution. Each point of this joint distribution has a probability mass assigned, which (under independence assumptions between the objectives) can be expressed by the marginal probabilities (the objective functions). Compared to the probabilistic solution it can be seen that imprecise probabilistic solutions are neither points, nor precise probability distributions. Instead, mass values are assigned to regions of the objective space. Again, assuming independence between the

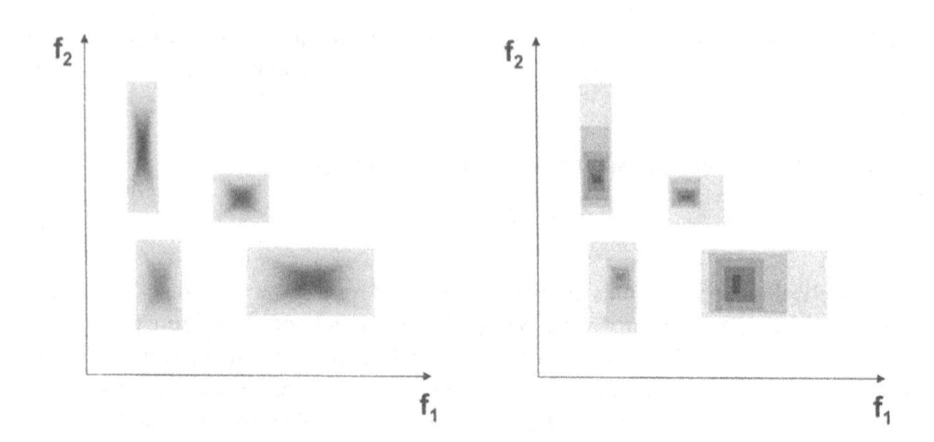

Fig. 6.2. Illustration of an objective space with probabilistic solutions (left) and imprecise solutions (right). In the probabilistic case, each point in $\mathbb{R}^{n_Y}$ has a precise probability (color gradient) to be the location of the solution. In the imprecise case, each region in $\mathbb{R}^{n_Y}$ has a belief and plausibility to be the location of the solution.

objective functions, the joint BPA $\widetilde{Y}$ can be expressed by the BPAs $\widetilde{Y}_1, ..., \widetilde{Y}_n$ of the single objectives. This is an even weaker assumption on the knowledge about the objective function that can be helpful if dealing with problems containing epistemic uncertainty.

Imprecise objective functions are not directly optimizable by common MOEAs, and the research on MOEA approaches for this class of objectives is still limited. There are four different ways to preprocess imprecise objective vectors prior to the passing to the MOEA, according to which the current research may be classified:

1. Averaging out all uncertainty and thus converting the imprecise solution to a point value.
2. Conversion of the imprecise solution to a sharp probability by assuming a distribution inside the focal elements, resulting in probabilistic objective values [159, 59, 9, 77, 108].
3. Comparing quantities such as the expected function values, resulting in interval-valued objectives [102, 110].
4. No preprocessing, direct utilization of the imprecise solution.

Depending on the above type of preprocessing, the applied MOEA must be modified to handle probabilistic (2), interval-valued (3) or imprecise probabilistic (4) solutions. The next sections start with a review on the few approaches on MOEAs for probabilistic and for interval-valued functions. As to the fourth category, the optimization of imprecise functions, there seems to be no approach that deals with the evolutionary optimization of imprecise solutions. In this thesis, a novel strategy of the fourth category which works directly on imprecise functions is introduced and evaluated on two examples.

6.1.2.1 Multi-objective Optimization in System Dependability

Commonly, dependability optimization contains several incommensurable objectives. On one hand there is a measure for the predicted dependability constituent (e. g. reliability or safety), which may either be a deterministic, probabilistic or otherwise uncertain value. On the other hand there are conflicting objectives, such as the price or weight of the solution, which make the selection of the maximally redundant system infeasible. There are several recent approaches in literature that cover multi-objective system dependability optimization. These approaches can be divided into two groups. The first use purely deterministic objective functions. The quantity *dependability* is represented by a sharp value without uncertainty.

In [161], the RAP for multi-state systems is solved using physical programming, an aggregation approach to multi-objective functions. They use two objectives, *expected system utility* and *system cost*. In [101], the objectives to be optimized are the *system survivability* and the *system costs*. A single-objective genetic algorithm with an aggregation approach is used to find a good solution. The approach described in [96] includes *reliability at a point in time*, *system cost* and *system weight*. Using an aggregation approach, single-objective tabu search is applied. In [30] the criterion is a *percentile of the system failure distribution*. Several runs are carried out for different percentiles to represent different risk preferences. [20], [106] and [107] use aggregation approaches which estimates the *net profit*, a difference between turnover and costs (acquisition, repair, downtime).

In the dependability prediction process, uncertainty may be explicitly represented. The second group of approaches model these uncertainties. The larger part of this group includes a measure for the uncertainty (e. g. variance) or a lower bound as a second optimization objective. In [29] a multi-objective formulation with three quantities of interest is used: Reliability at a given time, *system cost* and *system weight*. The reliability itself is uncertain and thus the quantity "reliability" is split into two objectives: *Expected reliability* to be maximized and *reliability variance* to be minimized. Similar in the objectives, approach [118] optimizes *expected reliability at a specific point in time* and *reliability variance* while putting weights and costs as constraints. In [28], the objectives are the expected reliability and the reliability variance. An aggregation method with changing weights is used for optimization. In [104], two quantities of interest, the expected system lifetime and system costs are optimized. The expected system lifetime is uncertain and given in bounds. The resulting objective functions are *expected system lifetime (lower bound)*, *expected system lifetime (upper bound)* and *system cost*. [108] apply a probabilistic dominance criterion for minimizing *system cost* and maximizing *system life*.

6.2 Pareto Evolutionary Algorithms for Multi-objective Optimization Problems

6.2.1 Evolutionary Algorithms: Overview and Terminology

Evolutionary algorithms are a class of population-based non-gradient methods whose rise in popularity began with the early works of [148], [136] and [76]. Introductions to single-objective evolutionary algorithms are e. g. [63] and [11]. Evolutionary algorithms try to mimic the principle of Darwinian evolution by natural selection. Each optimization parameter x_i is encoded by a "gene" in an "appropriate representation", e. g. a real number, integer or bit string. The kind of "appropriate representation" varies heavily from application to application and may have a strong influence on the optimization progress.

The evolutionary algorithms terminology is related to biological evolution. To prevent misunderstandings, a short introduction to the terms used is given. Evolutionary algorithms are optimization algorithms inspired by nature's strategy to create optimal life forms for a given environment. Evolution can be regarded as an optimization progress. Individuals who are well adapted to their environment survive and generate descendants who transport successful genetic material through time. This well-known principle is called selection. Selection is solely a phenotypic process. Offspring is normally generated through reproduction of two individuals that mix their genomes via crossover. At some point of the evolutionary process, random undirected mutations, changes in the genome of a single individual can occur. Mutations introduce new genetic material which enriches the variety in the population. Evolutionary algorithms make use of this principle to solve optimization problems (figure 6.3). The algorithm maintains a population (set) of individuals (e. g. representing possible system designs) which are the subject of an artificial evolutionary process. An individual consists of a genotype (decision vector) and a phenotype (objective vector). A decision vector is referred to as genome and a single value as gene. Starting point of each optimization run is an initial population of randomly created individuals (initialization). The initialization process is divided into generation steps consisting of selection, crossover, mutation and evaluation phases. Each generation, parenting individuals from the old population are selected for

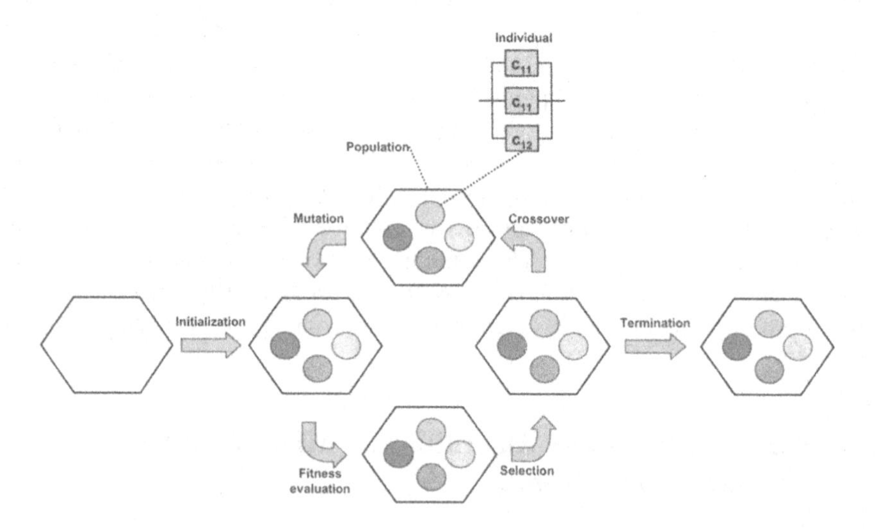

Fig. 6.3. Operating principle of a genetic algorithm

crossover and mutation. The newly created individuals that form the new generation are evaluated by the objective function and the next generation step starts. If a stop condition (e. g. number of generations exceeded) holds, then the algorithm ends, else the evolutionary cycle continues.

The MOEA terminology distinguishes between aggregating approaches and Pareto techniques (also a priori and a posteriori optimization). A priori methods require knowledge about user preferences beforehand. A multi-valued utility function is constructed from these preferences, which combines several objectives to one real-valued function. Single-criterion optimization algorithms are then used to find optimal solutions respective to the utility function. On the other hand, a posteriori methods do not require explicit user preferences at the beginning. Solely based on the Pareto dominance concept, a solution can only dominate another solution if it is at least equal in all objectives and better in one. The solution set consists of an approximation of the Pareto set. This set of solutions (figure 6.1), which most often has a fixed size is presented to the decision-maker for a posteriori selection of one or more optimal solutions. For two reasons, the solution set is only an approximation to the real Pareto optimal set. If the problem is very complex, solutions might be dominated by unfound optimal ones and therefore not be Pareto optimal. But second, the true Pareto optimal set may contain a large (or even infinite) number of solutions but the decision maker requires only a small set. The algorithm can then only present and pursue a small proportion of the true optimal set. Two important requirements have to be fulfilled by Pareto algorithms. On one hand, the optimization algorithm must be able to converge sufficiently fast towards the Pareto front. On the other hand, the optimization must preserve maximal diversity of the presented solutions. Almost all Pareto MOEAs therefore contain a diversity (or niching, anti-crowding) operator.

6.2.2 An Evolutionary Algorithm for Multi-objective Optimization under Uncertainty

In this subsection, it is discussed how a MOEA can be modified to be capable of solving imprecise probabilistic optimization problems. This MOEA, which is based on NSGA-II [40] and SPEA2 [184] implements new dominance criteria in the selection process and for maintaining a nondominated repository. Furthermore a new niching strategy is added, which helps to maximize the spread of the solutions even under imprecision.

Algorithm 6.1 shows the steps in the main function of the proposed MOEA. After an initialization phase where the initial generation *Pop* is created, the main loop begins. In the elitism step, the mating pool *Mat* is created. A predefined number *el* of individuals are randomly selected from the repository with maximal size *rsize*. This reintroduction of currently nondominated solutions (elitism) allows successful individuals to re-enter the evolutionary process in the hope for successful offspring.

The larger rest of the mating pool is filled with new individuals. In the selection phase, two individuals are randomly drawn from the old population. With a probability of $P_c \in [0,1]$, these individuals are recombined using single-point crossover. In the mutation step, each gene is mutated with a probability $P_m \in [0,1]$. The mutation adds a random number from a Gaussian distribution to the corresponding element. The new individuals are evaluated and included in the mating pool. This procedure is reiterated

Listing 6.1. The generic MOEA

```
1  ALGORITHM GMOEA (el ,psize ,msize ,Pc ,Pm)
2  Description
3     Generic MOEA
4  Definitions & declarations
5     Receives:
6        Elitism number el
7        Population size psize
8        Mating pool size msize
9        Crossover parameter Pc
10       Mutation parameter Pm
11    Returns: Rep
12    Solution set: Pop,Mat,Rep
13 Core
14    Initialize Pop
15    Evaluate Pop
16    Rep=Update Repository(∅,Pop)
17    WHILE Stop condition not met
18       Mat=Select el individuals randomly from Rep
19       WHILE |Mat| < msize
20          [ind1,ind2]= Select(Pop)
21          [ind1,ind2]= Crossover(ind1,ind2,Pc)
22          [ind1,ind2]= Mutation(ind1,ind2,Pm)
23          [ind1,ind2]= Evaluation(ind1,ind2)
24          Mat = Mat ∪ ind1
25          IF (|Mat| < msize)
26             Mat = Mat ∪ ind2
27          END IF
28       END WHILE
29       Mat=NSort(Mat)
30       Rep=Update Repository(Rep,Mat)
31       Pop=Mat[1:psize] \\the first psize individuals form the new
             generation
32    END WHILE
33    RETURN Rep
34    END ALGORITHM
```

until the mating pool is filled. The complexity of NSGA-II is described in [41]. It is $O(|mat|^2 \cdot n_Y + |mat| \cdot |f|)$ per generation. $|f|$ describes the complexity of the objective function that is unknown but usually large. The number of generations is defined by the stopping criteria or directly by the user.

The nondominated sorting approach is used in the selection phase. It has proven to be an effective selection criterion in the algorithms NSGA [158] and NSGA-II [40] and has since then been implemented in various other strategies. In order to create the new population, individuals in the mating pool *Mat* are sorted (figure 6.4). The first *psize* individuals in the sorted pool form the new population, representing the natural selection process.

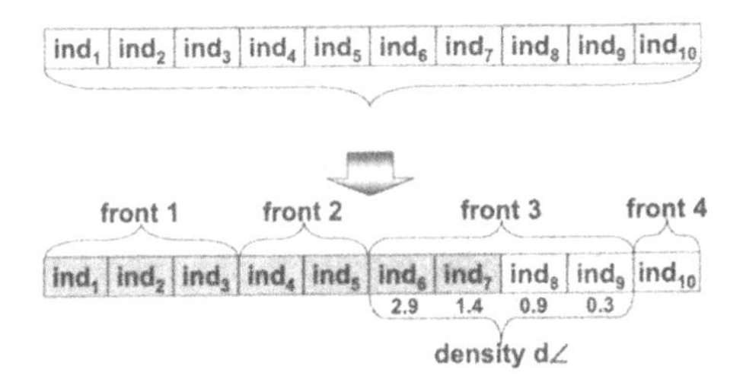

Fig. 6.4. Nondominated sorting process. Individuals are first sorted according to the nondominated front they belong too. The second sorting criterion is the density operator $d\angle$.

Listing 6.2. Algorithm for obtaining nondominated fronts with uncertain objective values

```
1   ALGORITHM NSort (Mat)
2   Description
3      Separates the solution set Mat into nondominated fronts Fr
4   Definitions & declarations
5      Receives: Solution set Mat
6      Returns: Fr
7      Vector of solution sets: Fr
8   Core
9      t = 1
10     WHILE Mat ≠ ∅
11       Fr[t] = {q ∈ Mat : ∄q' ∈ Mat : q' ≻ q} //≻ being the applied dominance
              relation
12       IF Fr[t] = ∅
13          Fr[t] = Mat
14       END IF
15       Mat = Mat\Fr[t]
16       t = t + 1
17     END WHILE
18     RETURN Fr
19  END ALGORITHM
```

As discussed, it is desirable to achieve two goals simultaneously during the optimization, convergence and diversity. Therefore, two sorting criteria are implemented. The first criterion $Fr(\widetilde{Y})$ is the nondominated front number of the individual $\widetilde{Y}$. If the front number is equal, the crowding measure is consulted for further ordering. A discussion on the applied crowding measure is given in section 6.4. The algorithm for determining Fr from [40] is given in listing 6.2.

After having generated the mating pool, the repository Rep is updated. All solutions in $Pop \cup Rep$, which are not dominated by any other solution in $Pop \cup Rep$ remain / are inserted in Rep. After having included all nondominated solutions, the size constraint is tested. While $|Rep| > rsize$, the individual with the highest density is deleted until the repository has the desired size.

6.3 Dominance Criteria on Imprecise Objective Functions

A MOEA constantly needs to compare solutions to progress in its optimization task. A dominance criterion is necessary to maintain the nondominated repository. Nondominated sorting requires a dominance criterion, too and if selection operators (e. g. tournament selection) are applied, they also require such a criterion. In the deterministic case, the Pareto dominance (eq. 6.3) works well. However, for imprecise probabilistic functions, there is no criterion that has been tested with MOEAs.

6.3.1 Probabilistic Dominance

If imprecise solutions are converted to probabilistic solutions, MOEAs that are able to deal with probabilistic functions can be applied. There is a small number of approaches based on probabilistic dominance, namely the works of [159], [77], [59] and [9]. A BPA can be transformed to a probability distribution by choosing an arbitrary distribution bounded by the belief and plausibility functions. With this step, all interval uncertainty is erased. One way is to replace each focal element by its mean, forming a discrete distribution. If this averaging procedure is the most adequate is not clear considering that a lot of effort was spent on the uncertainty model. If the uncertainty was reducible by simple averaging, the objective function would not be imprecise probabilistic. Secondly, a BPA encloses a lot of distributions with various shapes and properties, and it remains a degree of arbitrariness which one to select for comparison.

In the deterministic case, $f_i(x) \geq f_i(x')$ is either true or false. If the ojectives are probabilistic functions $\widehat{f}_i(x)$, a probability $P(\widehat{f}_i(x) \geq \widehat{f}_i(x')) \in [0,1]$ can be given and obtained as:

$$P(\widehat{f}_i(x) \geq \widehat{f}_i(x')) = \int_{-\infty}^{\infty} P(\widehat{f}_i(x) = y_1) \cdot P(\widehat{f}_i(x') \leq y_1) dy_1 \tag{6.6}$$

$$= \int_{-\infty}^{\infty} \int_{-\infty}^{y_1} P(\widehat{f}_i(x) = y_1) \cdot P(\widehat{f}_i(x') = y_2) dy_2 dy_1 \tag{6.7}$$

In [59] a probabilistic dominance criterion on a set of solutions is used. Even if no single point in a set $X_S \in X$ dominates x, many of them may almost do so. Therefore, the expected number of solutions in X_S dominating x, $E(X_S \succ_P x)$ is given as:

$$E(X_S \succ_P x) = \sum_{x' \in X_S} P(x' \succ_P x) \tag{6.8}$$

Alpha-dominance ($\succ_\alpha$) of a solution by a solution set is given, if the expected number of dominating solutions in X_S surpasses a threshold α.

$$X_S \succ_\alpha x :\Leftrightarrow E(X_S \succ_P x) \geq \alpha \tag{6.9}$$

Based on this alpha-dominance, solutions are in-/excluded into/from the repository. The probability of dominance was obtained assuming independent objectives:

$$P(x' \succ_P x) = \prod_{i=1}^{n} P(f_i(x') > f_i(x)) \tag{6.10}$$

It is not necessarily a good assumption to suppose independence between the objectives. In fact, it is much more logical to assume that there could be a strong negative dependence, as multi-objective optimization is applied in case of conflicting (negative dependent) objectives. If this dependence is not modeled explicitly, a criterion based on the dominance probability in each dimension can be applied. The probabilistic dominance relation $\succ_{Pr}$ [108] is an extension of this concept, based on the probability of dominance in each criterion:

$$x \succ_{Pr} x' :\Leftrightarrow \begin{cases} \forall i = 1, ..., n_Y : P(\widehat{f_i}(x) \geq \widehat{f_i}(x')) > p_{min} \\ \exists i = 1, ..., n_Y : P(\widehat{f_i}(x) > \widehat{f_i}(x')) > p_{min} \end{cases} \tag{6.11}$$

$p_{min} \in [0.5, 1[$ denotes a threshold of confidence that is needed to accept the statement "$\widehat{f_i}(x) \geq \widehat{f_i}(x')$". The larger p_{min} is chosen, the more the comparisons result in "neither better nor worse". This choice may be adequate if $\widehat{f}(x)$ is uncertain or potentially erroneous and can be interpreted as a margin for the dominance criterion. The MOEA then may choose according to user preferences, diversity in the solution set or other criteria.

6.3.2 Imprecise Probabilistic Dominance

If the dominance criterion from eq. 6.11 is transferred to the case of imprecise solutions, some modifications are necessary. Given imprecision, $Bel/Pl(\widetilde{f_i}(x) \geq \widetilde{f_i}(x')) \in [0, 1]$ bound the exact dominance probability. While transferring the concept of probabilistic dominance to the imprecise probabilistic case, it is feasible to put a threshold value on $Bel(\widetilde{f_i}(x) \geq \widetilde{f_i}(x'))$. Therefore, imprecise probabilistic dominance $\succ_{DS}$ can be defined as:

$$x \succ_{DS} x' :\Leftrightarrow \begin{cases} \forall i = 1, ..., n_Y : Bel(\widetilde{f_i}(x) \geq \widetilde{f_i}(x')) > p_{Bel} \\ \exists i = 1, ..., n_Y : Bel(\widetilde{f_i}(x) > \widetilde{f_i}(x')) > p_{Bel} \\ \forall i = 1, ..., n_Y : Bel(\widetilde{f_i}(x) > \widetilde{f_i}(x')) \geq Bel(\widetilde{f_i}(x') > \widetilde{f_i}(x)) \end{cases} \tag{6.12}$$

In contrast to the probabilistic case, $Bel(\widetilde{f_i}(x) \geq \widetilde{f_i}(x')) < 0.5$ does not imply $Bel(\widetilde{f_i}(x) < \widetilde{f_i}(x')) > 0.5$. Therefore, a much smaller $p_{Bel} \in [0, 1]$ may be chosen. The threshold value p_{Bel} represents the amount of belief necessary to accept that one solution is better than another. If $p_{Bel} = 0$, $x \succ_{DS} x'$ as long as $\forall i = 1, ..., n_Y : Bel(\widetilde{f_i}(x) > \widetilde{f_i}(x')) > Bel(\widetilde{f_i}(x') \geq \widetilde{f_i}(x))$. Otherwise, there must be at least a certain amount of belief for each objective to support this statement.

6.4 Density Estimation for Imprecise Solution Sets

In the case of deterministic solutions, almost all MOEAs include a density measure $d(y)$ to obtain a diverse solution set. The developed MOEA uses a density measure in the nondominated sorting (figure 6.4) as the second criterion if the front numbers are equal. Popular density measures are the bounding box approach (NSGA-II, [40]), the k-nearest neighborhood (SPEA2, [184]) and the adaptive hypercube method (MOPSO,

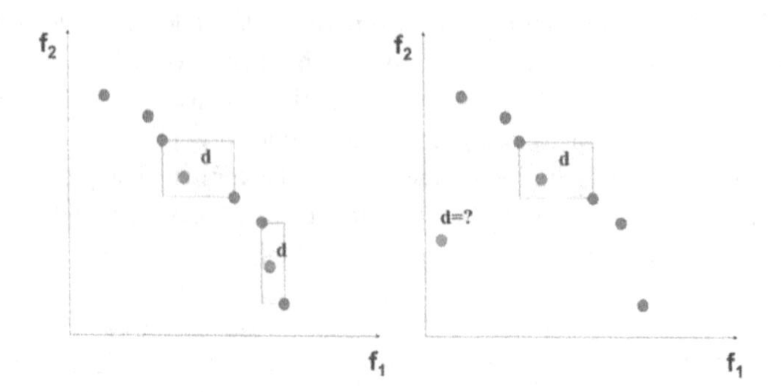

Fig. 6.5. Bounding boxes for density estimation. Left: Only Pareto Optimal solutions. The bounding boxes represent a good density estimate. Right: The measure is not applicable for dominated solutions, no bounding box exists.

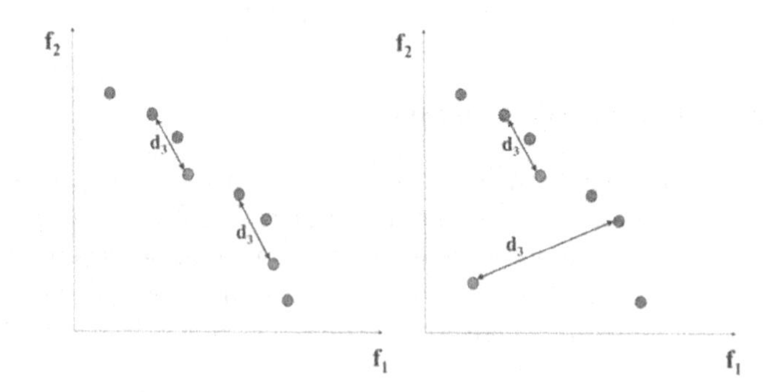

Fig. 6.6. 3-nearest neighborhood. Left: Only Pareto optimal solutions. The density measure performs well. Right: Dominated $\check{y}$ values are included. Being far from the nondominated set, these solutions have a high d-value and thus are rewarded for being dominated.

[27], DOPS [8, 75]). In the bounding box approach, $d(y)$ is the volume of the rectangular hypercube around y, constrained by the closest individuals (figure 6.5, left). Solutions on the "corners" of the front (a minimal/maximal value in one dimension) receive a maximal/infinite $d(y)$-value. The k-nearest neighborhood method defines $d(y)$ as the distance to the k-nearest neighbor. If $d_1(y),...,d_{n-1}(y)$ are distances from y to all other solutions in ascending order, $d(y) = d_k(y)$ (figure 6.6, left). Adaptive hypercube approaches divide the objective space in a set of disjunct hypercubes. The amount of solutions covering the same cube is used as a density measure.

In imprecise optimization, a way to define a fast density operator for the imprecise case is to convert all solutions to precise ones. This can be done by taking the "mean of the mean" $\check{y}$ of a solution as the point for density estimations:

$$\check{y} = \frac{Bel(E(\widetilde{Y})) + Pl(E(\widetilde{Y}))}{2} \tag{6.13}$$

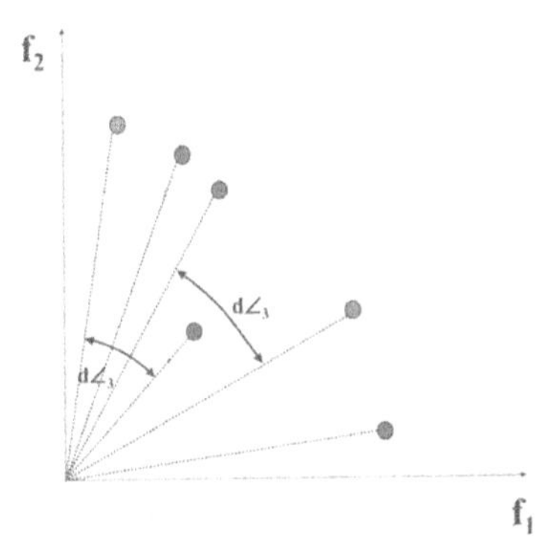

Fig. 6.7. 3-nearest angle. Even for dominated $\check{y}$-values, the 3-nearest angle provides a good density measure.

In this case, all uncertainty is "averaged out". It could be argued that uncertainty needs to be preserved in density estimation, too. For the sake of computational speed-up, the uncertainty in $\widetilde{Y}$ is neglected in this second criterion, as the determination of the first sorting criterion (front number) is already preserving uncertainty. With this sharp value, the $d(\check{y})$-value of a solution can be obtained. Even in a nondominated set of solutions it may be the case that some $\check{y}$ values are dominated (e. g. a nondominated solution with a large variance but a low $\check{y}$-value). Both the bounding box and the k-nearest neighborhood approach are only applicable for nondominated sets. If dominated solutions are included, the bounding box approach ceases to work, which can be seen in the example shown in figure 6.5 (right). It is not for all cases possible to create a box around a dominated solution. The k-nearest neighbor method is applicable, but its results are not useful. Solutions with $\check{y}$-values that are very far from the nondominated front are "rewarded" with a very high $d(\check{y})$-value and would be considered as much more valuable than other solutions. For the same reasons, the adaptive hypercubes method is not applicable. This problem can be solved by using not the distance, but the angular difference in relation to a reference point. The angular difference is the angle between the solutions:

$$d\angle(\check{y}, \check{y}') := \arccos\left(\frac{\check{y} \cdot \check{y}'}{|\check{y}| \cdot |\check{y}'|}\right) \tag{6.14}$$

The proposed algorithm utilizes a k-nearest neighbor approach on the angular distances with $d\angle_k(y)$ being the k-nearest angle. The angular difference can be considered as a projection of the solution set on the unit circle and solutions with dominated $\check{y}$ values are not "rewarded". Figure 6.7 illustrates the approach on an exemplary set with dominated and nondominated $\check{y}$-values.

6.5 Illustrative Examples

This section demonstrates the design space specification method and the proposed MOEA on a dependability optimization problem under uncertainty. The first subsection

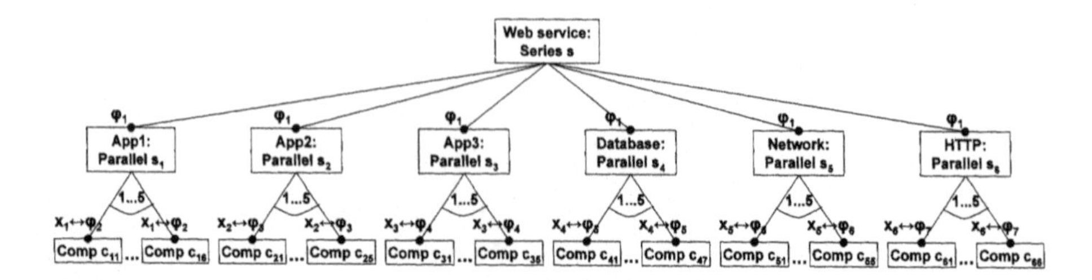

Fig. 6.8. RAP formulation of a redundant web service as a feature model (Ex1). The decision variables corresponding to features ($x_i \leftrightarrow \varphi_j$) are drawn in the figure. The series node "web service" has a mandatory feature set with each feature matching a parallel block (App1, App2, App3, Data base, Network, HTTP) in the RAP.

offers an overview on several studies regarding system dependability optimization. Then, the proposed design screening method is applied to two illustrative examples. While the first example (Ex1) is a RAP, modeling a web service, the second (Ex2) utilizes additional features and the resulting design space is much more complex. On both examples, design screenings are carried out and the resulting systems are shown. The illustrative example includes two objective functions: life distribution and system cost. The system is considered to be nonrepairable. Component life distributions may follow an arbitrary distribution. Objective $\widetilde{f_1}(x)$ is a random variable, the system life $\phi(s(x))$:

$$\widetilde{f_1}(x) = \phi(s(x)) \tag{6.15}$$

The objective $\widetilde{f_1}$ should be maximized. As the focus of this example is on the selection methodology and not on the system simulation itself, a simple cost function was chosen. The second objective function $\widetilde{f_2}(x)$ is the negated sum of all component costs:

$$\widetilde{f_2}(x) = B - \sum_{c \in s(x)} CS(c) \tag{6.16}$$

$CS(c)$ denotes the cost of component c, B is a constant (the budget) that is introduced to turn $\widetilde{f_2}$ into a maximisation objective.

6.5.1 RAP

The first design problem which is solved is a RAP with 6 parallel blocks (figure 6.8), modeling a web service with three different application servers which is obtained from [172]. A series system consisting of HTTP server, data base server, network connection and three application servers (App1, App2, App3) composes the web service. The goal is to find the optimal hardware configuration (component selection) for guaranteeing a high dependability with acceptable costs.

For each web service subsystem $s_1, \ldots, s_6$, 1,...,5 parallel hardware configurations from table 6.1 could be selected. All component life distributions were sampled from a Weibull function with different characteristic lifetime and shape parameter intervals

Table 6.1. Component data table (life parameters, costs) for Ex1 and Ex2

C_1				C_2				C_3			
Name	**Cost**	**α**	**β**	**Name**	**Cost**	**α**	**β**	**Name**	**Cost**	**α**	**β**
c_{11}	[1500,1800]	[10000,12000]	[1.2,1.3]	c_{21}	[3750,4350]	[60000,65000]	1	c_{31}	[4050,4500]	[42000,45000]	[1.4,1.4]
c_{12}	[1800,2100]	[14000,18000]	[1.6,1.7]	c_{22}	[4500,4800]	[80000,82000]	1	c_{32}	[3600,3900]	[45000,50000]	[1.8,1.8]
c_{13}	[2250,2400]	[18000,22000]	[1.1,1.2]	c_{23}	[5250,5550]	[100000,100000]	1	c_{33}	[5250,5700]	[48000,52000]	[1.3,1.3]
c_{14}	[2700,2700]	[22000,25000]	[1.5,1.6]	c_{24}	[6000,6750]	[120000,150000]	1	c_{34}	[2550,2700]	[51000,55000]	[1.7,1.7]
c_{15}	[3000,3750]	[26000,28000]	[1.4,1.5]	c_{25}	[6750,7200]	[140000,150000]	1	c_{35}	[4500,4500]	[54000,56000]	[1.4,1.4]
c_{16}	[3000,3300]	[30000,40000]	[1.5,1.6]								

C_4				C_5				C_6			
Name	**Cost**	**α**	**β**	**Name**	**Cost**	**α**	**β**	**Name**	**Cost**	**α**	**β**
c_{41}	[750,900]	[15000,17000]	1	c_{51}	[3450,3750]	[25000,28000]	[2.1,2.2]	c_{61}	[5250,5700]	[45000,48000]	[1.2,1.4]
c_{42}	[1500,1800]	[26000,29000]	1	c_{52}	[3600,4800]	[28000,32000]	[2.3,2.5]	c_{62}	[5700,6300]	[48000,48000]	[1.4,1.6]
c_{43}	[2250,2700]	[30000,33000]	1	c_{53}	[4650,5100]	[35000,41000]	[1.5,1.5]	c_{63}	[6300,6750]	[51000,55000]	[1.1,1.2]
c_{44}	[3000,3150]	[34000,40000]	1	c_{54}	[5250,5700]	[32000,33000]	[1.8,1.9]	c_{64}	[6900,7500]	[54000,59000]	[1,1.1]
c_{45}	[3300,3600]	[38000,38000]	1	c_{55}	[7650,8250]	[18000,20000]	[2,2.1]	c_{65}	[7800,8250]	[57000,58000]	[1.2,1.4]
c_{46}	[6000,6300]	[42000,45000]	1					c_{66}	[12000,12600]	[60000,64000]	[1.5,1.6]
c_{47}	[9000,10500]	[46000,49000]	1								

α, β using ADF sampling [162]. According to the feature model, the number of realizations per feature set are $(1,461,251,251,791,251,461)$ resulting in the decision vector (neglecting feature sets without variability):

$$x = \begin{pmatrix} x_1 \in \{1,...,461\} \\ x_2 \in \{1,...,251\} \\ \vdots \\ x_n \in \{1,...,461\} \end{pmatrix} \tag{6.17}$$

$|D| \approx 2.66 \cdot 10^{15}$ possible system realizations are contained in this design space, far beyond the possibility for total enumeration. The MOEA was run for 200 generations with a population size of 10 and a mating pool size of 10. This resulted in a total of 2,000 system evaluations. The repository size was restricted to 10 solutions. Crossover probability was set to 0.9, mutation probability to 0.1.

Figure 6.9 shows the result set of the optimization run. The imprecise probabilistic solution $\widetilde{f}(x)$ is expressed by the expected value, the 5th and the 95th percentile. As the objectives are imprecise distributions, all quantities are (multidimensional) intervals. The 10 points contained in the repository form a well-spread Pareto front. The solution set contains solutions with a high system life, but also solutions with low costs and trade-off solutions with medium costs and system life.

In figure 6.10, the optimization progress is shown. The expected values of the solutions contained in the initial repository (red), after 100 generations (blue) and after 200 generations (black) are given. In the initialization phase, where solutions are generated randomly, individuals are far from the optimal designs. After 100 generations, a solution set which is already well-spread has emerged. The final solution set improves the solution quality and helps spread the solutions further, offering a wide range of nondominated trade-off solutions to the decision-maker.

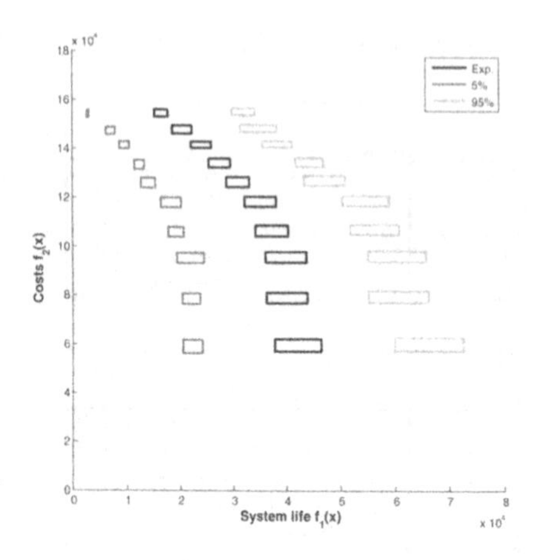

Fig. 6.9. Results for Ex1 (expected values, 5% and 95% quantiles). The solutions are well-spread over the whole objective space.

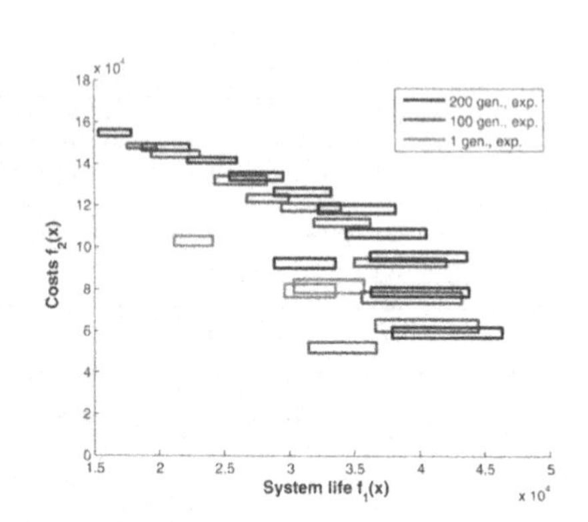

Fig. 6.10. Study on the optimization progress (Ex1). Plotted is the repository after initialization, 100 generations and 200 generations.

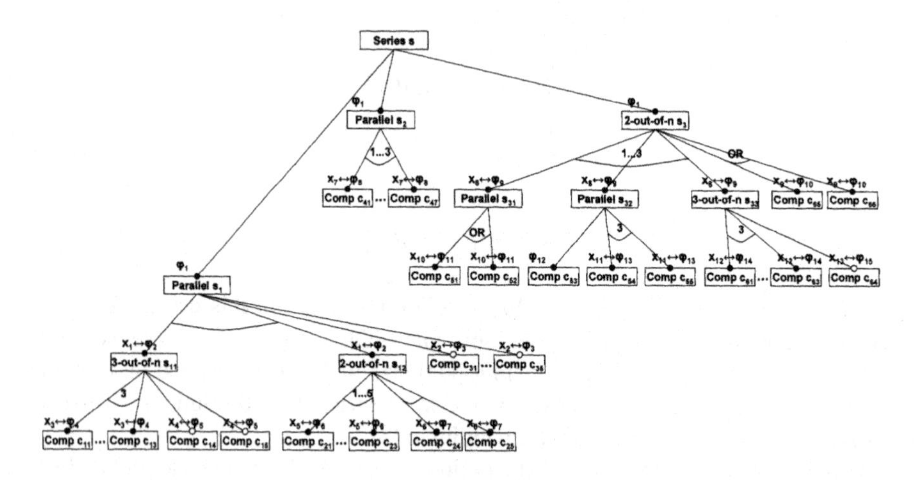

Fig. 6.11. Feature model with decision variables corresponding to features ($x_i \leftrightarrow \varphi_j$), Ex2

6.5.2 Complex Design Space

The second example (Ex2) is a feature model with a higher complexity (figure 6.11). However, the system under consideration is smaller, with only three subsystems in series at the root level. A series system with three mandatory blocks (two parallel and one 2-out-of-n) is the concept node. The first parallel system s_1 consists of either s_{11} or s_{12}. Other parallel components $c_{31},...,c_{35}$ are optional. s_{11} and s_{12} are itself complex systems. For example, s_{11} needs at least 3 components from type $c_{11},...,c_{13}$ plus

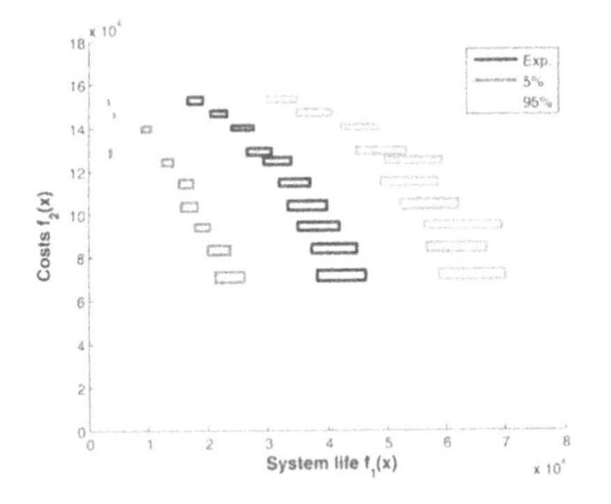

Fig. 6.12. Results for Ex2 (expected values, 5% and 95% quantiles). The solutions are cover a wide range of the objective space.

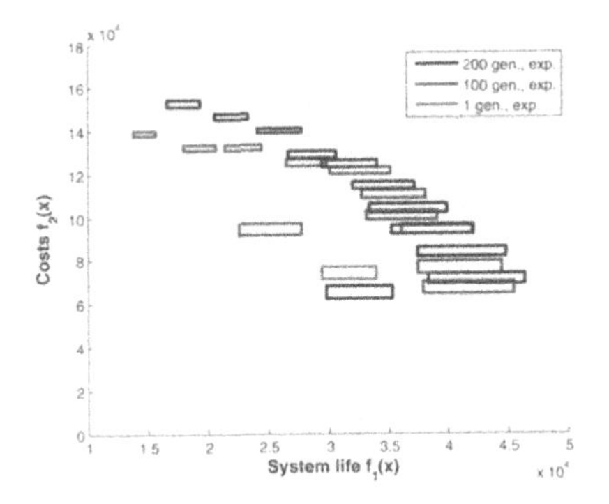

Fig. 6.13. Study on the optimization progress (Ex2). Plotted is the repository after initialization, 100 generations and 200 generations.

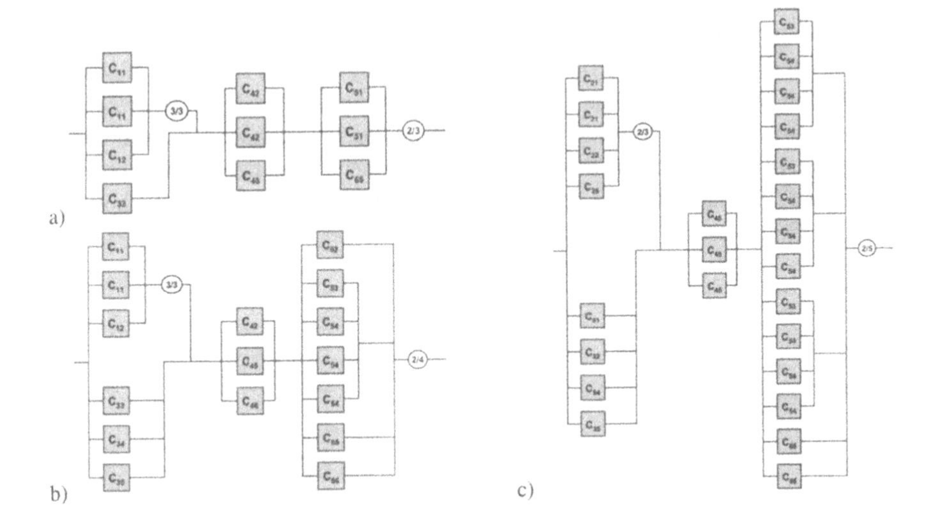

Fig. 6.14. Result systems after 200 generations (Ex2). a) Cost-efficient solution; a low amount of components has been used. b) Solution with a good trade-off. c) Highly redundant solution.

optional components c_{14} and c_{15}. The component cost and dependability parameters were inherited from the first example.

The MOEA was run with the same settings as in Ex1. The size of the decision space $|D| = 4.96 \cdot 10^7$ was much smaller than the example Ex1. The stronger restrictions that the user imposes with this feature model reduce the number of possible realizations and simplify the search process. The resulting node cardinalities were (1,2,32,10,4,55,2, 119,19,3,3,1,4,10,2). It can be observed, that the result set diversity (figure 6.12) is similar to Ex1, providing systems for various possible preferences of the decision-maker.

Figure 6.13 shows that this problem seemed more difficult to resolve. Especially in the area of low-budget systems with low costs and comparably low system life, there is a large difference between the 100th generation and the 200th generation. Three systems from the result set are shown in figure 6.14. System a) is more cost-efficient with a low amount of redundancy. System b) is a solution with a trade-off between high costs and high system life. System c) is a highly reliable system with a large amount of redundancy.

6.6 Conclusion

This chapter proposes a novel way to tackle the optimization of system dependability. Several aspects were integrated to enhance the practical usability of this optimization task for EDS. In conceptual design, where user preferences are not exactly known, a set of Pareto optimal system designs for further studies is desired. Thus, a MOEA for probabilistic functions based on a dominance relation for imprecise probabilistic objectives was developed to screen the design space for this set. With the new dominance criterion and a density operator based on angles, it is possible to optimize imprecise probabilistic functions without the indirect comparison of means or medians. The optimization method was applied in both problem formulations to find a set of systems that fulfill the requirement of the user. Solutions with varying trade-offs between cost and dependability and various degrees of uncertainty were obtained. A fruitful starting point of future research may be the investigation of other Pareto-based optimization algorithms on DST functions. The approaches described in [35] and [17], which include a representation of partial knowledge about user preferences are notably promising.

7 Case Study

This chapter illustrates the application of DST and MOEAs for quantifying and optimizing the uncertainty in a fault tree case study from the automotive field. The object of interest, an automatic transmission system under development emanates from the ZF AS Tronic family. The study represents the contribution of the University of Duisburg-Essen to the ESReDA (European Safety, Reliability & Data Association, [51]) workgroup on uncertainty analysis. This workgroup which unites industrial and research institutions (e. g. EADS, Électricité de France, JRC Ispra & Petten) aims at developing guidelines on uncertainty modeling in industrial practice based on a common framework developed from exemplary case studies. The chapter is structured as follows. After presenting the background of the study and introducing the IEC 61508 (section 7.1), a more detailed presentation on the system and the corresponding fault tree is given in section 7.2 and section 7.3. Section 7.4 introduces the IEC 61508 according to which safety requirements of the ATM are specified and shows how to integrate system and safety requirements in the ESReDA uncertainty analysis framework. Sections 7.5 and 7.6 describe how DST can help to check if the safety requirements are fulfilled. In sections 7.7 and 7.8, possible improvements regarding safety are screened using feature models and evolutionary algorithms. The obtained result set, containing several promising solutions is analyzed and could be postprocessed by the responsible decision-maker.

7.1 Background

The IEC 61508 is a generic industrial standard which is implemented e. g. in railway [44] and medical engineering [79]. It helps to manage the requirements on functional safety of electronic, programmable but also of mechatronic development projects. In the German automotive industry, the IEC 61508 plays a key role for safety analysis. This is a very recent development as the standard had been rejected by the "Verband der Automobilindustrie" (VDA), the German Association of the Automotive industry in 2002. However, in 2004, the VDA had withdrawn this decision and introduced the IEC 61508 to replace all other functional safety standards [147].

In conformance with this standard, Safety Integrity Levels (SIL 1-4) are used to define quantitative system safety requirements. Uncertainty analyses of the predicted

P. Limbourg: Dependability Modelling under Uncertainty, SCI 148, pp. 107–121, 2008.
springerlink.com

system safety provide both a robust way to demonstrate that the system complies with the target failure measure and an indicator for possible violations of these targets.

The IEC 61508 proposes quantitative prediction methods such as the fault tree analysis to obtain probabilities for critical events [156]. Quantitative dependability and safety prediction in an early design stage needs to deal with uncertainties from various sources. Data is obtained from expert estimates and models have a low degree of detail and may be only an inaccurate description of the real failure behavior. Uncertainty preserving models are therefore necessary for predicting the compliance with a SIL in an EDS. Precise outputs would neglect the high uncertainty of the input data. In this study, a fault tree analysis of an automatic transmission from the ZF AS Tronic product line is the base on which the developed methods are applied. The study aims at finding out if the system complies with the target failure measure of SIL 2 and which paths could be followed to ensure compliance even under EDS uncertainty is carried out. Based on the analysis of the existing system, several design alternatives are screened for a better trade-off between failure probability and costs.

7.2 System under Investigation

The system under investigation is a currently developed member of the ZF AS Tronic series (figure 7.1). ZF AS Tronic automatic transmissions have been developed especially for trucks with EDC (electronic diesel control) engines and CAN (controller area network) communications [183]. The transmission system, combining mechanic transmission technology with modern electronics is a true mechatronic system according to definition 2.1. The integrated modular design simplifies both installation and maintenance and allows flexible configuration for different scopes of function. The transmission can be equipped with one or more clutch-dependent power take-offs, even after it has been installed in the truck. The outputs can be shifted independently of each other. A speed-dependent power take-off system to drive auxiliary steering pumps is also available. ZF AS Tronic handles gear selection, clutch and shifting maneuvers. The

Fig. 7.1. Automatic transmission from the ZF AS Tronic product line: (1) Transmission actuator, (2) Gearbox, (3) Clutch actuator

12-speed gearbox shifts electropneumatically. Engine power is always transmitted at a near-optimal level. The ZF-MissionSoft driving program keeps the motor at an efficient engine speed. The driver can correct the automatic gear selection or switch to manual operation at any time and set the gear using the touch lever. The main application fields of this ATM are buses, trucks and other special-purpose vehicles.

7.3 Fault Tree Model

Being only a subsystem in the power train that is again a subsystem in a modern car, it is difficult to specify exactly which failure modes may lead to a safety-critical situation for the driver. However, certain conditions can be identified as critical, such as a failure of the clutch system or an uncovered failure in the gearbox electronics. In this study, the detail level of the fault tree (figure 7.2) is quite low. Subsystems are mainly not separated deep down to part level.

The system is composed of three different subsystems. The clutch system mainly contains mechanical components. The gearbox with a set of mechanic and hydraulic components will include several sensors that may detect failures or near-failure conditions. The gearbox electronics being the interface to the driver is modeled as a separate subsystem. It is necessary to mention that while the ATM modularization is the real one, expert estimates and additional component choices are assumed and do not reflect the real failure behavior of the AS Tronic.

As described in section 2.4, the probability of a component $i = 1,...,n$ to fail per working hour is described by a failure probability $p_i \in [0,1]$ and p represents the input parameter vector of the predefined system model. In an EDS, uncertainties perturb the knowledge on p. The uncertainty study will therefore illustrate how to form an uncertainty model around p and project this uncertainty onto the output, the system failure probability p_s, representing the variable of interest for the case study.

7.4 Quantifying Reliability According to the IEC 61508 and the ESReDA Uncertainty Analysis Framework

According to the IEC 61508, it has to be shown for a safety-critical function that the probability of a critical failure is below a certain threshold (target failure measure) to comply with a certain SIL. Systems are classified according to their usage frequency into high demand and low demand. For systems with high demand such as the transmission under investigation, table 7.1 shows the corresponding target failure measure (failure probability per hour of service). The IEC 61508 provides intervals for these target failure measures. This motivates the use of uncertainty preserving methods such as [140]. If it can be shown that the predicted failure measure p_s is possibly lower than the lower target failure measure bound and with a high degree of confidence below the higher target failure measure bound, then the safety arguments are much stronger.

The IEC 61508 plays a major role in safety assessment. Although originally intended for the use with electronic / programmable systems, it has rapidly gained importance in mechatronics, too. However, its role is rather to use some parts of the validation method

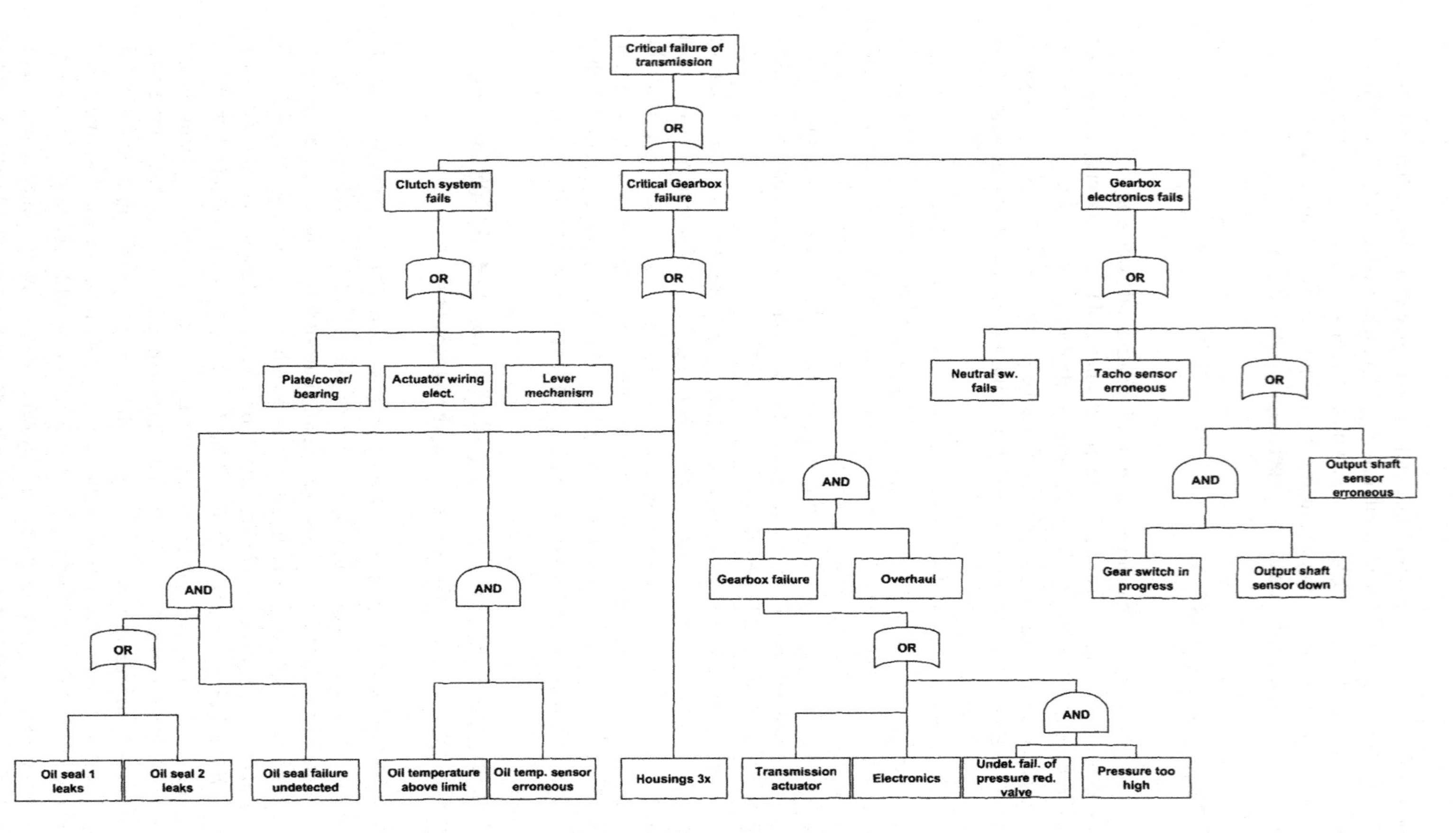

Fig. 7.2. Early design stage fault tree of an automatic transmission

Table 7.1. Tolerated target failure measure for safety-critical functions (high demand or continuous mode) [78]

SIL	Target failure measure
4	[1E-9, 1E-8]
3	[1E-8, 1E-7]
2	[1E-7, 1E-6]
1	[1E-6, 1E-5]

than a one-to-one implementation. The SILs are a common term for reliability and safety requirements even in this case, but e. g. some aspects on redundant functionality following from a certain SIL are not possible to achieve if mechanical components are involved.

The IEC 61508 moreover defines limits on the SIL that can be reached by a system executing the safety function depending on data quality and fault tolerance. (Sub-) systems are classified based on the quality of available data. A (sub-) system is of the (preferable) type A if:

1. the behavior of all components in case of failure is well-defined and
2. the behavior of the subsystem in case of failure is well-defined and
3. reliable failure data from field experience exists.

A system is of type B if at least one of the criteria 1-3 is not fulfilled. In an EDS, type B is considerably much more likely to occur. Even if criteria 1 & 2 are well-defined, it is unlikely that there exists enough data to fulfill criterion 3. Due to the recursivity of this definition, systems containing one type B subsystem have to be considered as type B, too. However, this has a strong impact on the safety requirements. Systems of type A can reach higher SILs than type B systems.

It may be desirable to be able to classify a system of type A. If criteria 1 and 2 are fulfilled, but exact failure data is missing, it may be reasonable to work with a conservative uncertainty model. If the uncertainty model can be considered as reliable (regarding criterion 3), the system may be treated as type A. Therefore, very conservative uncertainty modeling such as implicitly done in DST can help to provide arguments for shifting the (sub-)system from type B to type A. In contrast to purely probabilistic uncertainty treatment, a DST approach enables the inclusion of uncertainty in form of intervals (section 3.9). This eases the communicability of the uncertainty study. While arguing that the system falls into type A, modeling the uncertainty of a critical variable in a probabilistic way needs to satisfy the question of the right choice of the distribution. Facing a skeptical observer, an answer may not be easy. In a scenario involving DST, such critical uncertainties can be captured in intervals or sets of probability distributions.

Figure 7.3 illustrates the integration of the study into the ESReDA uncertainty analysis framework. The core of the analysis, the pre-existing model is a system fault tree, describing the system failure probability as a function of component failure probabilities. The component failure probabilities may be uncertain and thus represent the uncertain model inputs. The chosen uncertainty model is a joint DST belief function on the

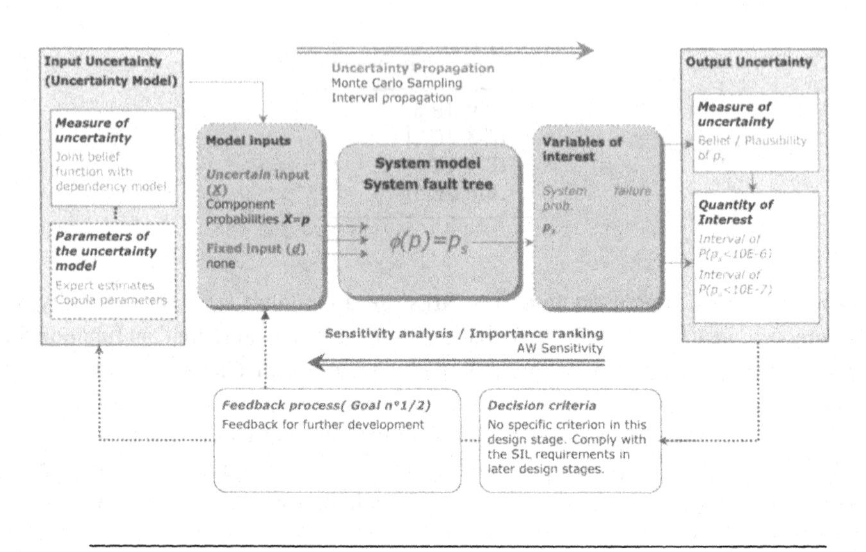

Fig. 7.3. Uncertainty analysis on the ATM failure probability. Integration into the ESReDA framework on uncertainty modeling.

component probabilities. The uncertainty is propagated onto the system failure probability, resulting in a measure of uncertainty on p_s. The main quantities of interest are the belief and plausibility to violate the SIL threshold. No formal decision criterion is implemented, as the goal of the study is rather design feedback than compliance testing.

7.5 Quantification of the Input Sources

As indicated, the advantage of DST is its flexibility to handle expert estimates. Therefore different ways of predicting failure probabilities are provided. The experts may predict p_i using several methods:

- Estimate a sharp value for p_i.
- Estimate an interval that contains p_i.
- Estimate mean μ and standard deviation σ of a normal distribution (censored to [0,1]) describing the uncertainty in p_i.

Experts are also allowed to provide several estimates on the same component that are aggregated using weighted mixing. Intervals and sharp values are represented as focal elements with mass 1. Estimates on μ and σ could be given as values or intervals. The specification of parameters of a probability distribution in DST describes a set of functions that each encloses the probability distribution. Sampling can be interpreted as propagating the estimated distribution parameters μ and σ together with a BPA of

Table 7.2. Expert estimates on the failure probability of different components

Basic Event	ID	Source	Failure prob./ h of service	Disc.
Plate/Cover/Bearing	1	Expert 1	[1E-8,3E-8]	Interval
		Expert 2	4E-8	Point
		Expert 3	μ=1E-7, σ=[1E-8,2E-8]	ODF
Actuator wiring & elect.	2	Expert 1	μ=[1E-8,2E-8], σ=1E-9	ODF
		Expert 2	4E-8	Point
		Expert 3	μ=2E-8, σ=1E-9	Conf
Lever mechanism	3	Expert 1	[0,1E-8]	Interval
Oil seal 1 leaks	4	Expert 1	[1E-7,3E-7]	Interval
		Expert 2	μ=3E-7, σ=[1E-7,2E-7]	ODF
		Expert 3	μ=[3E-7,5E-7], σ=1E-8	ODF
Oil seal 2 leaks	5	Expert 1	[1E-7,3E-7]	Interval
		Expert 3	μ=2E-7, σ=1E-8	Conf
Oil seal failure undetected	6	Expert 2	1E-9	Point
Oil temp. above limit	7	Expert 2	[0.01,0.03]	Interval
Oil temp. sensor erroneous	8	Expert 1	3E-7	Point
		Expert 3	μ=4E-7, σ=1E-8	Conf
Housings 3x	9	Expert 2	0	Point
		Expert 3	[0,1E-9]	Interval
Transmission actuator	10	Expert 1	[1E-8,3E-8]	Interval
		Expert 2	μ=7E-8, σ=3E-9	ODF
		Expert 3	[2E-8,4E-8]	Interval
Electronics	11	Expert 2	μ=3E-6, σ=5E-7	ODF
Undet. fail. of pressure red. valve	12	Expert 1	[7E-8,1.5E-7]	Interval
		Expert 2	[1E-7,2E-7]	Interval
		Expert 3	4E-7	Point
Pressure too high	13	Expert 3	0.02	Point
Overhaul fails	14	Expert 1	[0.003,0.005]	Interval
		Expert 2	μ=[1E-3,2E-3], σ=2E-4	Conf
		Expert 3	μ=3E-3, σ=[1E-4,2E-4]	Conf
Neutral sw. fails	15	Expert 1	6E-8	Point
		Expert 2	[0,1E-7]	Interval
Tacho sensor erroneous	16	Expert 2	5E-7	Point
		Expert 3	[2E-7,4.5E-7]	Interval
Gear switch in progress	17	Expert 3	0.04	Point
Output shaft sensor down	18	Expert 2	μ=2E-7, σ=4E-8	ODF
Output shaft sensor erroneous	19	Expert 1	μ=3E-8, σ=1E-8	ODF
		Expert 2	μ=4E-8, σ=1E-8	ODF
		Expert 3	μ=7E-8, σ=2E-8	ODF

probabilities m_p describing the sampling strategy. The BPA $m_{i,j}$ estimated on component i by expert j is derived as:

$$m_{i,j} := F^{-1}_{Gauss}(m_p, \mu, \sigma) \tag{7.1}$$

a) 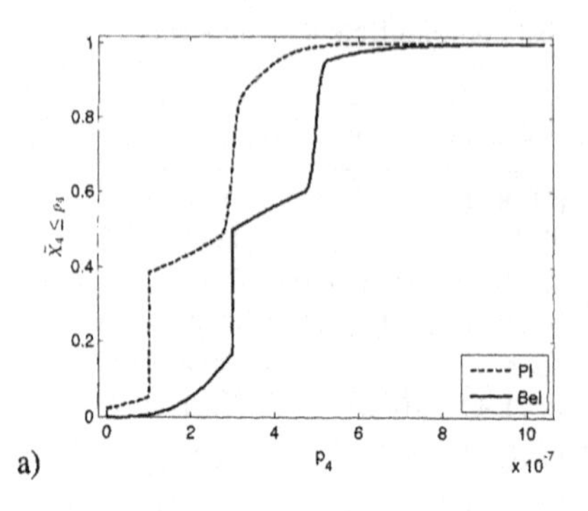

b) 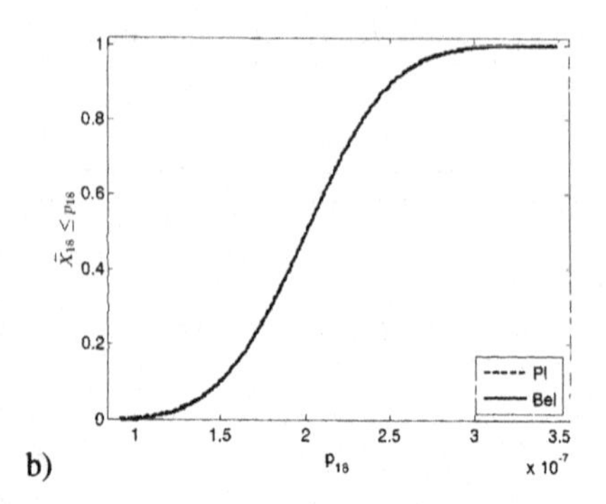

Fig. 7.4. Visualization of the belief and plausibility functions of the aggregated estimates of basic events: a) oil seal 1 leaks and b) output shaft sensor down

The estimates were converted by sampling the inverse CDF of the distribution. Two different sampling techniques were supported, the outer discretization (ODF, [162]) method and confidence sampling (Conf, [112]). This approach is a research study with no real data available. However, fault tree and expert estimates have been derived and validated in close collaboration with a reliability and safety expert from ZF Friedrichshafen AG.

To illustrate the estimation process, three experts (Expert 1, Expert 2 and Expert 3), who provide estimates on the failure probability were assumed. The values are idealized and do not reflect the real failure probabilities. Table 7.2 shows these estimates on the basic fault tree events. While on some components like component 6 (oil seal failure undetected) there is only one estimate, component 10 (transmission actuator) has several estimates, which are aggregated to one marginal distribution. Figure 7.4 illustrates three different BPAs of input components. It can be seen in figure 7.4 a) that interval estimates lead to an instantaneous increase in the Bel / Pl functions. Figure 7.4 b) is a pure distribution estimate without any interval uncertainty, and thus Bel and Pl collapse to a CDF.

7.6 Practical Implementation Characteristics and Results of the Uncertainty Study

Fault tree models have a very low amount of computation. Thus, 1,000,000 samples of the joint BPA were propagated through the system function using Monte Carlo sampling. All calculations performed and all plots shown were generated using the imprecise probability toolbox for MATLAB. This free & open source tool, developed in the context of this thesis is available at [103]. Figure 7.5 shows the prediction on belief and plausibility of the system failure probability and gives some characteristics of the resulting BPA. The vertical lines represent the upper and lower failure probability required by SIL 2. It can be seen that the stricter SIL probability is surpassed. However, the lower threshold 1E-6 is not surpassed with both high belief and plausibility. With a plausi-

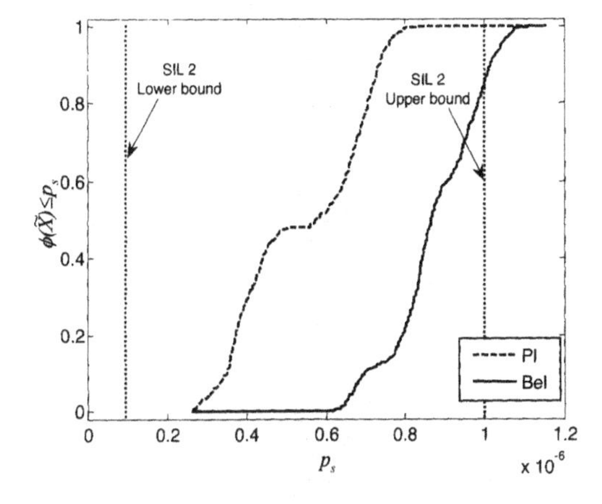

$E(p_s)$	[5.2E-7 8.1E-7]
Med(p_s)	[5.5E-7 8.1E-7]
$Q_{95}(p_s)$	[7.5E-7 1.0E-6]
Bel/Pl(p_s¡1E-6)	0.92 / 1
Bel/Pl(p_s¡1E-7)	0 / 0
GH	11.4485
AW	3.2794E-7

Fig. 7.5. Results of the fault tree analysis: Belief and Plausibility of the system failure probability p_s, illustrated SIL 2 compliance thresholds, statistics on the BPA of p_s

bility of 100% and a belief of 92%, the SIL compliance is reached. The uncertainty contributed by the interval width is quite high. If the belief of compliance should also reach 100%, then interval uncertainty must be reduced. Expectation value and median are approximately equal. These values, which average over the uncertainty modeled by distributions but not over the interval uncertainty can be given in bounds. The bounds are narrower for the median than for the expectation value.

While dealing with a second order DST uncertainty treatment, it is necessary to communicate the meaning of the uncertainty expressed by the system output. The uncertainty is not aleatory (aleatory uncertainty is implicitly represented by the probabilistic fault tree). It stems from expert uncertainty and conflict between estimates. It can not simply be "averaged out" by producing more than one system. This important difference had to be pointed out if results were to be used in a decision process. Interval uncertainty can only be reduced if interval uncertainty on the input parameters is reduced, which is only possible by refining the uncertainty model with further information. As this normally involves additional effort (e. g. collecting more data, iterative expert elicitation rounds), the analyst may be interested in the area where such refinement is most effective. For this purpose, a sensitivity analysis using the *AW* measure was performed. The results show clearly where to collect more evidence. As can be seen in figure 7.6, reductions of the nonspecifity on the tacho sensor failure (ID 16) and the failure of the neutral switch (ID 15) may lead to the largest reduction of nonspecifity in the model output. The plate/cover/bearing failure (ID 1) seems to contribute highly to the general nonspecifity, too. Other components such as oil seal leakage (IDs 4,5,6) contribute less or zero nonspecifity. This may either be the case because the amount of information on the component is already high or because the component plays an insignificant role in the system model. It is important to note that sensitivity values do not add up to 100% but are to be treated as independent values, each describing the expected reduction of the overall uncertainty given maximal information about one component. From the sensitivity results, a strategy for further refinement of the dependability prediction can be

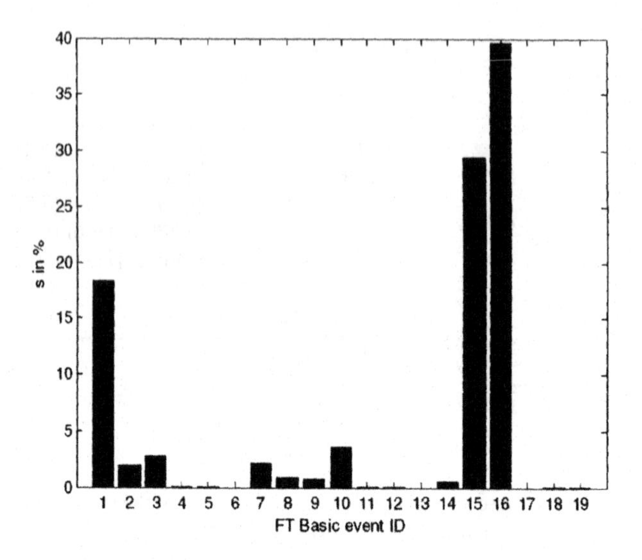

Fig. 7.6. Expected reduction of the overall nonspecifity (*AW* measure) given perfect information of one component

deduced. A concurrent approach is to try to find a design, which is more reliable and safe even under the uncertainties involved in the prediction. The next two sections cover this design screening process in more detail.

7.7 Specifying Design Alternatives

After obtaining dependability figures for the ATM, the question on how to improve them arises. Sensitivity indices are a valuable indicator for deducing a test strategy for further improvement of the dependability prediction. One goal of early dependability analysis is to aid in the system optimization process. This can be achieved by iterative improvement and re-evaluation of the system until a system with adequate safety emerges. As shown in chapters 5 and 6, automatic design screening is an alternative way. If this method is applied, it must first be specified which alternatives on the component and subsystem level are available. The cross product of all alternative choices on component and subsystem level form the set of all possible system design alternatives.

In the presented example, plates and housings could be produced using several material strengths, both with different failure probabilities and different costs. For the lever mechanism, an alternative design is possible. Sensors, standard COTS (commercial off-the-shelf) components could be obtained as different makes from different manufacturers. The following list shows all component variations considered:

Plate/Cover/Bearing: 5 different material strengths.
Lever mechanism: 2 alternative designs.
Oil temp. sensor: 5 different types.
Housings: 5 different material strengths.
Electronics: 4 different variants.

Table 7.3. Alternative component choices with costs and failure probability

Component	ID	Type	Failure prob./ h of service	Disc.	Costs
Plate/Cover/Bearing	1	Original (Type 1)	see tab. 7.2	see tab. 7.2	10
Lever mechanism	3	Original (Type 1)	see tab. 7.2	see tab. 7.2	100
Oil temp. sensor	8	Original (Type 1)	see tab. 7.2	see tab. 7.2	7
Housings 3x	9	Original (Type 1)	see tab. 7.2	see tab. 7.2	35
Electronics	11	Original (Type 1)	see tab. 7.2	see tab. 7.2	28
Tacho sensor	16	Original (Type 1)	see tab. 7.2	see tab. 7.2	8
Output shaft sensor	18	Original (Type 1)	see tab. 7.2	see tab. 7.2	4
Plate/Cover/Bearing	1	Type 2	μ=5E-8, σ=[1E-8,2E-8]	ODF	8
Plate/Cover/Bearing	1	Type 3	μ=9E-8, σ=[1E-8,2E-8]	ODF	7.5
Plate/Cover/Bearing	1	Type 4	μ=1.2E-7, σ=[1E-8,2E-8]	ODF	7
Plate/Cover/Bearing	1	Type 5	μ=1.8E-7, σ=[1E-8,2E-8]	ODF	6.5
Lever mechanism	3	Type 2	[1E-9,1.2E-8]	Interval	98
Lever mechanism	3	Type 3	[5E-9,1E-8]	Interval	95
Oil temp. sensor	8	Type 2	μ=2E-7, σ=1E-8	Conf	9
Oil temp. sensor	8	Type 3	μ=1.5E-7, σ=1.5E-8	Conf	9
Oil temp. sensor	8	Type 4	μ=6E-7, σ=1E-8	Conf	6
Oil temp. sensor	8	Type 5	μ=4E-7, σ=3E-8	Conf	6.5
Housings 3x	9	Type 2	[0,1E-9]	Interval	32
Housings 3x	9	Type 3	[0,3E-9]	Interval	30
Housings 3x	9	Type 4	[2E-9,7E-9]	Interval	28
Housings 3x	9	Type 5	[5E-9,1.1E-8]	Interval	27
Electronics	11	Type 2	μ=1E-6, σ=4E-7	ODF	35
Electronics	11	Type 3	μ=3E-6, σ=2E-7	ODF	30
Electronics	11	Type 4	μ=4E-6, σ=7E-7	ODF	22
Tacho sensor	16	Type 2	1E-7	Point	10
Tacho sensor	16	Type 3	2E-7	Point	9
Tacho sensor	16	Type 4	7E-7	Point	4
Output shaft sensor	18	Type 2	μ=2E-8, σ=4E-9	ODF	7
Output shaft sensor	18	Type 3	μ=4E-8, σ=4E-9	ODF	6
Output shaft sensor	18	Type 4	μ=6E-8, σ=4E-9	ODF	5
Watchdog	20	Type 1	μ=1E-6, σ=4E-7	ODF	3.5
Watchdog	20	Type 2	μ=1.5E-6, σ=4E-7	ODF	3
Watchdog	20	Type 3	μ=2E-6, σ=4E-7	ODF	1.5

Tacho sensor: 4 different types.
Output shaft sensor: 4 different types.

For each component failure probability, one expert estimate is given. A cost number is attached to each component, reflecting the development, integration and production costs (abstract units) that will come along with its selection. Costs are also uncertain, because it is never sure how much effort the integration of a component will actually require. To reflect the uncertainty, a triangular distribution ranging from 95% to 105% of the estimated costs was assumed. An overview over all component choices is given in table 7.3. Only component choices that represent a trade-off at component level were included. If a component had both a higher failure probability and higher costs, it can be a priori excluded from the screening process.

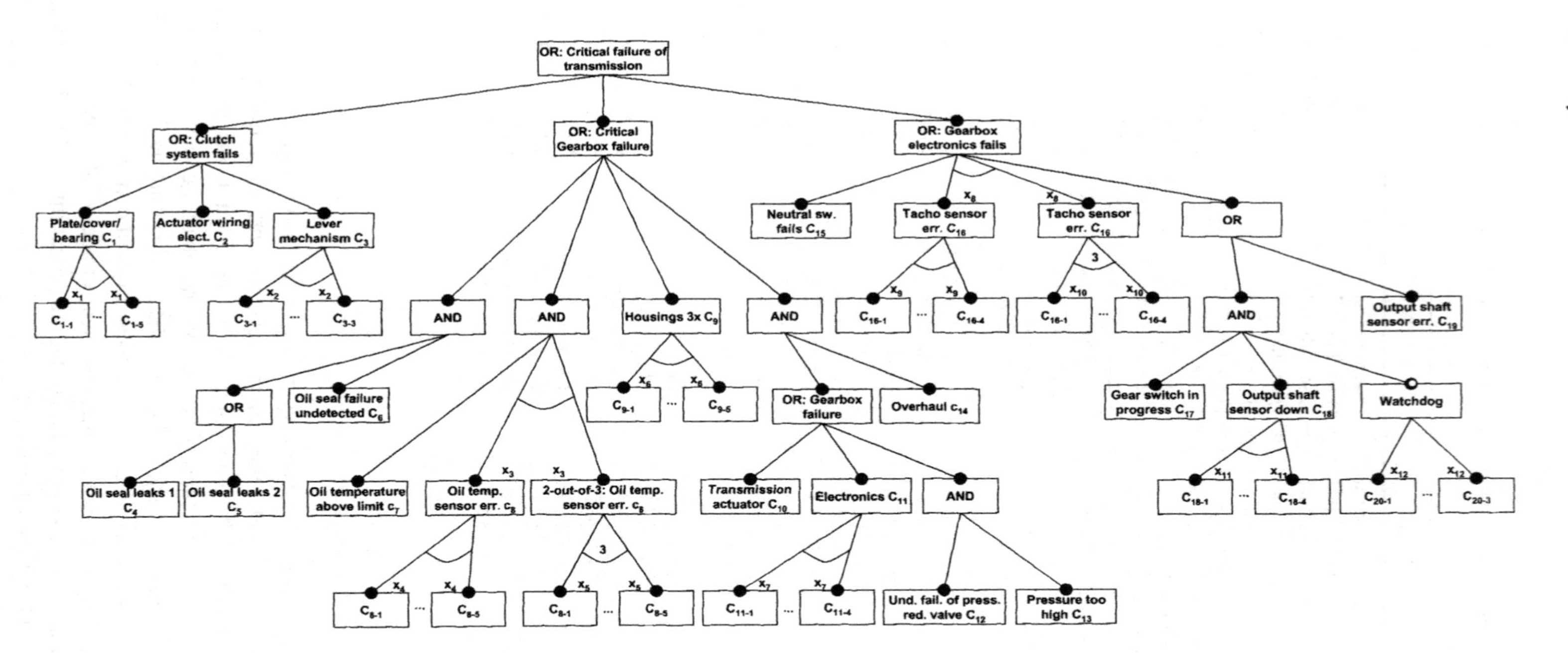

Fig. 7.7. ATM feature model. The feature model represents all component choices and structural variants of the ATM.

In addition to the component variations, there are points in the architecture, where safety measures such as redundancy could be implemented. To prevent erroneous sensor signals, triple structural redundancy with a 2-out-of-3 voting algorithm [48] could be included. The resulting subsystem fails only if two sensors malfunction. This possibility was modeled for the oil- and the tacho sensor. A mixture of several sensor types in a redundant subsystem is possible. The outage of the shaft sensor could be detected by an optional watchdog. Watchdogs are components which check the status of critical subsystems / components and react (e. g. by warning) if a component fails [122]. While watchdogs can detect outages, they do not necessarily discover erroneous sensor signals, which makes this approach less tolerant to faults than redundancy. For the shaft sensor, a watchdog was included in the design space. In an overview, the system structure design alternatives are:

Oil temp. sensor: 2-out-of-3 redundancy.
Tacho sensor: 2-out-of-3 redundancy.
Output shaft sensor: Optional watchdog.

Both the component-level and architectural changes can be subsumed in a feature model as shown in figure 7.7. The redundancy choice was formulated as a dimension choosing either one component or a 2-out-of-3 subsystem. The watchdog is represented as an optional feature. Component choices are modeled similar to the example in section 5.5. The relatively small amount of possible alternatives on the component level results in 1,152,000 design alternatives to screen for an optimal trade-off between dependability and costs. Since all components are on their own a trade-off between low costs and low failure probability, it is not easy to deduce which systems are the optimal ones.

7.8 Optimizing System Reliability

The developed feature model can be screened by an optimization algorithm such as presented in chapter 6. For this study, the MOEA was run for 200 generations with population and mating pool size of 10. This results in 2,000 screened systems for a run. Mutation probability was set to 0.1, crossover probability to 0.9. The repository, yielding the results of the screening process was limited to 10. A visualization of the resulting solution set in the objective space is shown in figure 7.8. Expected value, 5th and 95th percentile are shown by red, blue and green boxes.

It can be seen that the Pareto front has at its lower end a set of interesting trade-off solutions, where with a relatively low amount of extra costs the failure probability can be significantly reduced (systems 2, 3, 7, 9 and 10). With an increase of the budget by about 15%, the failure probability can be decreased by about 10% (solutions 2 and 10, expected values). Further increase of the budget has only little effect on the failure probability (systems 5, 6, 4, 8 and 1). Figure 7.9 shows a study on the optimization progress. The initial repository (red) does not contain any of the interesting trade-offs. Solutions after 100 generations (1,000 screened systems) are already in a near-optimal region and form a diverse front in the objective space. The final results after 200 generations slightly decrease the failure probability in the lower region, showing further progress and discovering better trade-off solutions.

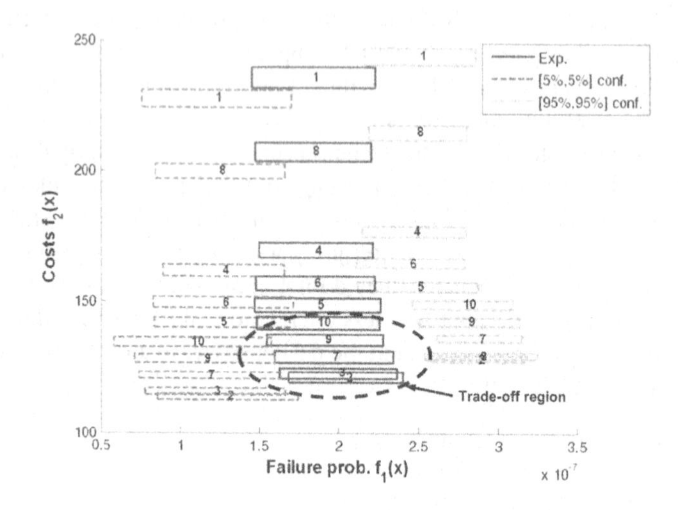

Fig. 7.8. Results for the ATM case study (expected values, 5% and 95% quantiles). The solutions show a set of good trade-offs in the lower region of the objective space (circle). Numbers inside the solutions indicate the corresponding system. It can be seen that with an increase of the budget by 15%, the failure probability can be decreased by about 10% (solutions 2 and 10).

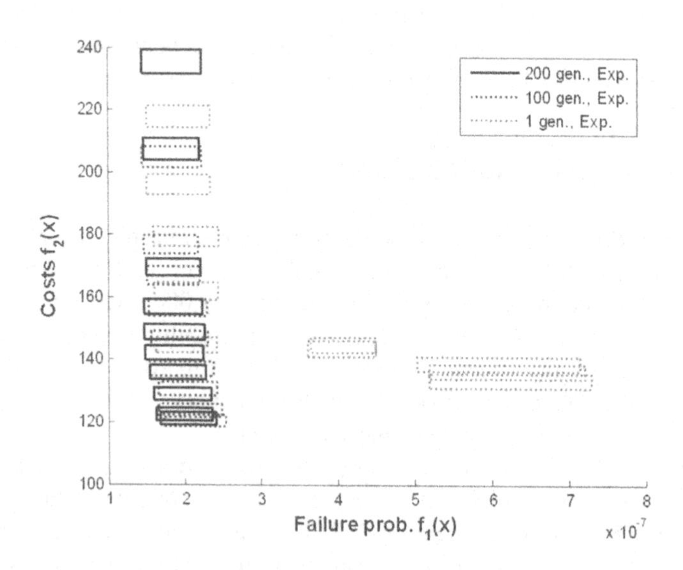

Fig. 7.9. Study on the optimization progress. Plotted is the repository after initialization, 100 generations and 200 generations. The repository contents after 100 generations already form a well-spread solution set. The final set (200 generations) shows further improvements in regions with lower failure probability.

Table 7.4. Three result systems

Feature	System 2	System 10	System 7
Plate/Cover/Bearing	Type 1	Type 1	Type 1
Lever mechanism	Type 3	Type 1	Type 3
Oil temp. sensor	Triple redundancy 3x Type 1	Triple redundancy 3x Type 1	Triple redundancy 3x Type 1
Housings	Type 1	Type 1	Type 1
Electronics	Type 4	Type 2	Type 4
Tacho sensor	Triple redundancy 3x Type 1	Triple redundancy 3x Type 1	Triple redundancy 2x Type 1, 1x Type 4
Output shaft sensor	Type 1	Type 2	Type 4
Shaft sensor watchdog	None	None	Type 2
$E(p_s)$	[1.72E-7,2.45E-7]	[1.51E-7,2.26E-7]	[1.63E-7,2.37E-7]
$Q_{95}(p_s)$	[2.66E-7,3.3E-7]	[2.43E-7,3.07E-7]	[2.55E-7,3.19E-7]
$E(C)$	[119,123]	[140,144]	[127,131]
$Q_{95}(C)$	[127,129]	[148,151]	[134,137]

From the results, one or several systems could be selected by the decision-maker for further analysis. Three promising ones are shown in table 7.4. Listed are the configurations for all variation points. Expected value and 95% quantile of the failure probability and the costs are given. System 2 realizes a system with low costs. No additional watchdog is included, but triple redundancy is implemented for the oil temperature sensor and the tacho sensor. System 10 involves higher costs but achieves a lower failure probability. System 7 is an intermediate case with a slightly higher probability of failure than system 10, but less costs. All three systems represent – on the base of the optimization run – near-optimal solutions which can be evaluated by engineers on their feasibility.

8 Summary, Conclusions and Outlook

8.1 Summary and Main Contributions

To meet the demand for integrated dependability assessment issued by mechatronic projects, dependability predictions especially for early design stages are to be developed and refined. These methods need to consider the various sources which introduce a large amount of uncertainty into the prediction. As a consequence, predictions must include uncertainty models. Beside probability theory, several other frameworks for representing and propagating uncertainty exist. As it is an open discussion whether or not the probabilistic representation is preferable to model epistemic uncertainty, other representations need to be evaluated.

In this work, a framework for dependability prediction and optimization in an early design stage based on Dempster-Shafer theory was presented. For allowing the transfer of DST to realistic dependability predictions, methodological fragments contained in various studies from other application areas and from theoretical works have been formed to a coherent methodology. While there have been works on DST for abstract dependability problems with a low model complexity, industrially relevant dependability studies are still conducted in a probabilistic setting. Thus, the activation of DST for realistic dependability studies can be seen as the main contribution of this thesis. Copulas were chosen to model dependence between belief variables and sensitivity analyses on uncertainty measures were used to measure the contribution of input variable uncertainty to the overall system uncertainty. To support the practitioner in his choice between a probabilistic and a DST setting, the advantages and drawbacks of DST (e. g. higher flexibility, higher computing requirements) are respectively discussed and it is outlined, how results in DST can be interpreted and communicated.

To support design-for-dependability in early design stages, a new way of specifying dependability optimization problems based on feature models is illustrated. The concept of design-screening, discovering a whole subset of candidate systems for further development in a single optimization run, was realized using feature models and multi-objective evolutionary algorithms. Feature models were used to describe sets of fault trees and reliability block diagrams, allowing complex constraints and dependencies between realizations. With the aid of an enumeration scheme, an interface for optimization

P. Limbourg: Dependability Modelling under Uncertainty, SCI 148, pp. 123–125, 2008.
springerlink.com

algorithms was created, representing the first use of feature models in a dependability screening context.

In conceptual design, where user preferences are not exactly known, a set of Pareto optimal system designs for further studies is preferable to a single solution. To screen the design space for the Pareto set even under uncertainty, a novel multi-objective evolutionary algorithms has been designed that is capable of handling several objective functions in a DST representation. Based on a dominance relation for imprecise probabilistic objectives and an angular distance density operator, DST functions can be optimized without the detour on expected values or medians. With this evolutionary algorithm, it is possible to find a set of systems with diverse trade-offs between cost and dependability under uncertainty.

To enhance the base data available for predictions in early design stages, a new way of reusing data from previous dependability analyses has been developed. By combining expert estimates on the similarity between old and new systems with the data base of a past project, the method is able to give robust quantitative predictions on the new system's dependability. Instead of defining a prediction algorithm a priori, Gaussian processes (Kriging) and neural networks, two methods from computational intelligence were used to regress the prediction function from training data. Thus, similarity prediction can be learned from examples, training cases provided by experts. In this way, flexibility is maintained and the prediction accuracy grows with the number of data available. The performance of the approach was evaluated on two test sets. The first, a scalable test suite for an arbitrary number of training cases, was developed as a generic benchmark. The second test set was based on real reliability data of four automatic transmissions. Neural networks and Gaussian processes were both able to predict the failure functions with acceptable accuracy. The Gaussian processes are more robust against input data that is strongly different from the examples and are therefore preferable if the training data quality is low. Results indicate that both learning approaches are suitable for CDF prediction in early design stages. Therefore, the method is especially useful if dependability data bases can be exploited or projects are developed as parts of larger product lines with several preceding products.

The presented methodology has been applied in an exemplary study, where the failure probability of an automatic transmission had to be predicted under uncertainty. From the results, violations of and conformance with SIL thresholds can be deduced under highly conservative uncertainty assumptions. A sensitivity analysis on a non-specifity measure was used to identify the components with the main contribution to the overall interval uncertainty. By means of feature models, a design space with several different component and subsystem structures was defined and successfully screened by the developed MOEA, resulting in a set of systems with good dependability-cost trade-offs. The study represents the contribution of the University of Duisburg-Essen to the ESReDA workgroup on uncertainty analysis and will be part of a set of guidelines on uncertainty modeling in industrial practice [50]. This can be interpreted as a partial validation on the industrial usability of the proposed methods, even if they have not been applied to a true industrial project yet. All tools for the representation and propagation of uncertainty have been released in an open source toolbox for MATLAB (Imprecise probability toolbox, [103]) to spread the usage of DST for uncertainty propagation.

8.2 Outlook

This thesis has shown that DST can be a convenient alternative to model system dependability. However, many questions on more advanced modeling techniques remain unanswered. It has been shown that variance-based sensitivity analysis in the area of DST is possible [58], but it has not been applied in a dependability context. Thus it would be tempting to include it in the toolkit of the early design stages dependability analyst for more detailed cause-consequence analyses in case of exceeding failure probabilities. Another step in this direction is the inclusion and interpretation of combined randomness-nonspecifity measures on the amount of uncertainty in belief functions. While there have been measures and algorithms proposed in [2] and [86] for discrete frames of discernment, it is not clear how to transfer these algorithms to real-valued BPAs [87, personal communication].

Other points which could be followed up in further research include the refinement of the similarity prediction method by investigating other learning approaches such as recurrent neural networks or genetic programming [7]. In case of larger data bases, similarity prediction can be combined with expert systems that propose similar components to the engineer which he can include in the estimation process. The most promising sources are data bases that are hosted and maintained by more than one company [142]. This opposes to todays classification of dependability data as highly confidential. However, this restriction may be relaxed if the value of a shared data pool for internal dependability analyses rises. With similarity prediction methods such as the proposed approach, old dependability data can be efficiently reused and may therefore become more important for the analyst.

The extension of the MOEA by including partial objective preferences [17, 35] for enhanced flexibility is also a promising way to go. With partial preferences, the user can give imprecise, partial information on the trade-off regions that are interesting, which may be included in the search process. Partial preferences can be considered as an intermediate approach between a priori and a posteriori (Pareto) optimization. In any case, the most tempting future work would be the long-term application of the prediction method in mechatronic projects, beginning in the early design stages. This is the only way to observe how uncertainty reduces in projects, and how good predictions match field estimates. However, such research involves an enormous amount of time and resources and thus could only be the starting point of a larger research project.

References

1. Abellan, J., Moral, S.: A non-specificity measure for convex sets of probability distributions. International Journal of Uncertainty, Fuzziness and Knowledge-Based Systems 8(3), 357–367 (2000)
2. Abellan, J., Moral, S.: Maximum of Entropy in Credal Classification. In: Bernard, J.M., Seidenfeld, T., Zaffalon, M. (eds.) ISIPTA 2003 - Third International Symposium on Imprecise Probability: Theories and Applications, pp. 1–15. Carleton Scientific, Lugano, Switzerland (2003)
3. Agarwal, H., Renaud, J.E., Preston, E.L., Padmanabhan, D.: Uncertainty quantification using evidence theory in multidisciplinary design optimization. Reliability Engineering and System Safety 85(1-3), 281–294 (2004)
4. Avizienis, A., Laprie, J.C., Randell, B., Landwehr, C.: Basic Concepts and Taxonomy of Dependable and Secure Computing. IEEE Transactions on Dependable and Secure Computing 1(1), 11–33 (2004)
5. Bae, H.R.: Uncertainty Quantification and Optimization of Structural Response using Evidence Theory. Phd thesis, Wright State University, Dayton, USA (2004)
6. Bae, H.R., Grandhi, R.V., Canfield, R.A.: An approximation approach for uncertainty quantification using evidence theory. Reliability Engineering and System Safety 86(3), 215–225 (2004)
7. Banzhaf, W., Nordin, P., Keller, R.E., Francone, F.: Genetic Programming - An Introduction; On Automatic Evolution of Computer Programs and its Applications. Morgan Kaufmann, San Francisco (1998)
8. Bartz-Beielstein, T., Limbourg, P., Mehnen, J., Schmitt, K., Parsopoulos, K.E., Vrahatis, M.N.: Particle swarm optimizers for Pareto optimization with enhanced archiving techniques. In: Sarker, R., Reynolds, R., Abbass, H., Tan, K.C., McKay, B., Essam, D., Gedeon, T. (eds.) IEEE Congress on Evolutionary Computation (CEC 2003), vol. 3, pp. 1780–1787. IEEE Press, Canberra, Australia (2003)
9. Basseur, M., Zitzler, E.: A Preliminary Study on Handling Uncertainty in Indicator-Based Multiobjective Optimization. In: EvoWorkshops 2006, Budapest, Hungary, pp. 727–739 (2006)
10. Baudrit, C., Hélias, A., Perrot, N.: Uncertainty analysis in food engineering involving imprecision and randomness. In: Cooman, G.d., Vejnarova, J., Zaffalon, M. (eds.) ISIPTA 2007 - Fifth International Symposium on Imprecise Probability: Theories and Applications, Prague, Czech Republic (2007)
11. Bäck, T.: Evolutionary algorithms in theory and practice: evolution strategies, evolutionary programming, genetic algorithms. Oxford Univ. Press, New York (1996)

12. Bedford, T.: The "Value of Information" in a Reliability Programme. In: Bedford, T., Cojazzi, G. (eds.) 27th ESReDA Seminar, Glasgow, United Kingdom, pp. 3–14 (2004)
13. Benz, S.: Eine Entwicklungsmethodik für sicherheitsrelevante Elektroniksysteme im Automobil (in German). In: Anheier, W. (ed.) 15. GI/ITG/GMM Workshop Testmethoden und Zuverlässigkeit von Schaltungen und Systeme, Timmendorfer Strand, Germany (2003)
14. Berleant, D., Zhang, J.: Bounding the Times to Failure of 2-Component Systems. IEEE Transaction on Reliability 53(4), 542–550 (2004)
15. Bertsche, B., Lechner, G.: Zuverlässigkeit im Fahrzeug- und Maschinenbau - Ermittlung von Bauteil- und System-Zuverlässigkeiten (in German), 3rd edn. Springer, Berlin (2004)
16. Bishop, C.M.: Neural Networks for Pattern Recognition. Oxford University Press, New York (1995)
17. Branke, J., Deb, K.: Integrating user preferences into evolutionary multi-objective optimization. In: Jin, Y. (ed.) Knowledge Incorporation in Evoluionary Computation. Studies in Fuzziness and Soft Computing, pp. 461–478. Springer, Berlin (2004)
18. Bundesrepublik Deutschland: V-Modell XT Dokumentation (in German) (2005), `http://www.kbst.bund.de`
19. Callan, R.: The Essence of Neural Networks. Prentice-Hall, Upper Saddle River (1999)
20. Cantoni, M., Marseguerra, M., Zio, E.: Genetic algorithms and Monte Carlo simulation for optimal plant design. Reliability Engineering and System Safety 68(1), 29–38 (2000)
21. Carreras, C., Walker, I.D.: Interval Methods for Fault-Tree Analyses in Robotics. IEEE Transactions on Reliability 50(1), 3–11 (2001)
22. Chern, M.S.: On the Computational Complexity of Reliability Redundancy Allocation in a Series System. Operations Research Letters 11(5), 309–315 (1992)
23. Chin, W.C., Ramachandran, V., Cho, C.W.: Evidence Sets Approach For Web Service Fault Diagnosis. Malaysian Journal Of Computer Science 13(1), 84–89 (2000)
24. Chiu, W.F.: Belief Functions Applied to Reliability Testing and Product Improvement. Phd thesis, Harvard University, Cambridge, USA (2004)
25. Citti, P., Delogu, M., Fontana, V., Ammaturo, M.: Reliability forecast of automotive systems based on soft computing techniques. In: Kolowrocki, K. (ed.) European Conference on Safety and Reliability – ESREL 2005, Balkema, Gdynia-Sopot-Gdansk, Poland, vol. 1, pp. 373–379 (2005)
26. Coello Coello, C.A.: The EMOO repository: a resource for doing research in evolutionary multiobjective optimization. IEEE Computational Intelligence Magazine 1(1), 37–45 (2006)
27. Coello Coello, C.A., Salazar Lechuga, M.: MOPSO: A Proposal for Multiple Objective Particle Swarm Optimization. In: Fogel, D., Atmar, W. (eds.) IEEE Congress on Evolutionary Computation (CEC 2002), vol. 2, pp. 1051–1056. IEEE Press, Honolulu (2002)
28. Coit, D., Jin, T., Wattanapongsakorn, N.: System Optimization Considering Component Reliability Estimation Uncertainty: A Multi-Criteria Approach. IEEE Transaction on Reliability 53(3), 369–380 (2004)
29. Coit, D.W., Baheranwala, F.: Solution of stochastic multi-objective system reliability design problems using genetic algorithms. In: Kolowrocki, K. (ed.) European Conference on Safety and Reliability – ESREL 2005, Balkema, Gdynia-Sopot-Gdansk, Poland, vol. 1, pp. 391–398 (2005)
30. Coit, D.W., Smith, A.E.: Genetic algorithm to maximize a lower-bound for system time-to-failure with uncertain component Weibull parameters. Computers and Industrial Engineering 41(4), 423–440 (2002)
31. Coolen, F.: On the use of imprecise probabilities in reliability. Quality and Reliability Engineering International 20, 193–202 (2004)
32. Cormen, T.H., Leiserson, C.E., Rivest, R.L., Stein, C.: Introduction to Algorithms, 2nd edn. MIT Press, Cambridge (2001)

33. Cox, D., Oakes, D.: Analysis of Survival Data. Chapman and Hall, London (1984)
34. Craig, K.: The Role of Computers in Mechatronics. Computing in Science and Engineering 5(2), 80–85 (2003)
35. Cvetkovic, D., Coello Coello, C.A.: Human Preferences and their Applications in Evolutionary Multi-Objective Optimisation. In: Jin, Y. (ed.) Knowledge Incorporation In Evolutionary Computation. Studies in Fuzziness and Soft Computing, pp. 479–503. Springer, Berlin (2004)
36. Czarnecki, K.: Generative Programming - Principles and Techniques of Software Engineering Based on Automated Configuration and Fragment-Based Component Models. Phd thesis, Technical University of Ilmenau, Ilmenau, Germany (1998)
37. Czarnecki, K.: Overview of Generative Software Development. In: Banâtre, J.-P., Fradet, P., Giavitto, J.-L., Michel, O. (eds.) UPP 2004. LNCS, vol. 3566, pp. 313–328. Springer, Heidelberg (2005)
38. Czarnecki, K., Eisenecker, U.W.: Generative Programming: Methods, Tools, and Applications. Addison-Wesley, Reading (2000)
39. Czarnecki, K., Kim, C.H.P.: Cardinality-Based Feature Modeling and Constraints: A Progress Report. In: OOPSLA 2005 International Workshop on Software Factories, San Diego, USA, pp. 211–220 (2006)
40. Deb, K., Agrawal, S., Pratab, A., Meyarivan, T.: A Fast Elitist Non-Dominated Sorting Genetic Algorithm for Multi-Objective Optimization: NSGA-II. In: Schoenauer, M., Deb, K., Rudolph, G., Yao, X., Lutton, E., Merelo, J.J., Schwefel, H.P. (eds.) Parallel Problem Solving from Nature VI Conference, Paris, France, pp. 849–858 (2000)
41. Deb, K., Pratap, A., Agarwal, S., Meyarivan, T.: A Fast and Elitist Multiobjective Genetic Algorithm: NSGA-II. IEEE Transactions on Evolutionary Computation 6(2), 182–197 (2002)
42. Dempster, A.: Upper and lower probabilities induced by a multivalued mapping. Annals of Math. Statistics 38, 325–339 (1967)
43. Denoeux, T.: Reasoning with imprecise belief structures. International Journal of Approximate Reasoning 20, 79–111 (1999)
44. DIN Deutsches Institut für Normung e.V: DIN EN 50128 - Railway applications - Communications, signalling and processing systems - Software for railway control and protection systems. Beuth, Berlin, Germany (2001)
45. DIN Deutsches Institut für Normung e.V: DIN IEC 61025 - Störungsbaumanalyse (in German). Beuth, Berlin, Germany (2004)
46. Démotier, S., Schön, W., Denoeux, T.: Risk Assessment Based on Weak Information using Belief Functions: A Case Study in Water Treatment. IEEE Transactions on Systems, Man and Cybernetics 36(3), 382–396 (2006)
47. Dubois, D., Prade, H.: On the combination of evidence in various mathematical frameworks. In: Flamm, J., Luisi, T. (eds.) Reliability Data Collection and Analysis, Eurocourses: Reliability and Risk Analysis, vol. 3, pp. 213–241. Springer, Netherland (1992)
48. Echtle, K.: Fehlertoleranzverfahren (in German). Springer, Berlin (1990)
49. Embrechts, P., Lindskog, F., McNeil, A.: Modelling dependence with copulas and applications to Risk Management. In: Rachev, S. (ed.) Handbook of Heavy Tailed Distributions in Finance, pp. 329–384. Elsevier, Amsterdam (2003)
50. de Rocquigny, E., Devictor, N., Tarantola, S. (eds.): Uncertainty in Industrial Practice: A Guide to Quantitative Uncertainty Management. Wiley, Chichester (2008)
51. European Safety, Reliability & Data Association: ESReDA Homepage (2007), `http://www.esreda.org/`
52. Eusgeld, I., Echtle, K., Kochs, H.D., Limbourg, P.: Automatic Design of Reliable Systems Consisting of Nano-Elements. In: 7th IEEE International Conference on Nanotechnology, Hong Kong, China (in press, 2007)

53. Eusgeld, I., Stoica, C.: Automatic Design Optimisation for Reliable Systems with Human Operators. In: Kolowrocki, K. (ed.) European Conference on Safety and Reliability – ESREL 2005, Balkema, Gdynia-Sopot-Gdansk, Poland, vol. 1, pp. 549–554 (2005)
54. Ferson, S., Ginzburg, L., Kreinovich, V., Nguyen, H., Starks, S.: Uncertainty in risk analysis: towards a general second-order approach combining interval, probabilistic, and fuzzy techniques. In: FUZZ-IEEE 2002. Proceedings of the 2002 IEEE International Conference on Fuzzy Systems, Honolulu, USA, vol. 2, pp. 1342–1347 (2002)
55. Ferson, S., Hajagos, J., Berleant, D., Zhang, J., Troy Tucker, W., Ginzburg, L., Oberkampf, W.: Dependence in Dempster-Shafer theory and probability bounds analysis. No. SAND2004-3072 in Sandia Report. Sandia National Laboratories, Albuquerque, USA (2004)
56. Ferson, S., Joslyn, C.A., Helton, J.C., Oberkampf, W.L., Sentz, K.: Summary from the epistemic uncertainty workshop: consensus amid diversity. Reliability Engineering and System Safety 85(1-3), 355–369 (2004)
57. Ferson, S., Kreinovich, V., Ginzburg, L., Myers, D.S., Sentz, K.: Constructing Probability Boxes and Dempster-Shafer Structures. No. SAND2002-4015 in Sandia Report. Sandia National Laboratories, Albuquerque, USA (2003)
58. Ferson, S., Tucker, W.T.: Sensitivity in Risk Analyses with Uncertain Numbers. No. SAND2006-2801 in Sandia Report. Sandia National Laboratories, Albuquerque, USA (2006)
59. Fieldsend, J., Everson, R.: Multi-objective Optimisation in the Presence of Uncertainty. In: IEEE Congress on Evolutionary Computation (CEC 2005), pp. 476–483. IEEE Press, Edinburgh (2005)
60. Finkelstein, M.: Simple Repairable Continuous State Systems of Continuous State Components. In: Mathematical Methods in Reliability (MMR 2004), Santa Fe, USA (2004)
61. Frank, M.J.: On the simultaneous associativity of F(x, y) and x + y - F(x, y). Aequationes Math. 19, 194–226 (1979)
62. Gausemeier, J.: Von der Mechatronik zur Selbstoptimierung (in German). In: 20th CAD-FEM Users Meeting 2002. Friedrichshafen, Germany (2002)
63. Goldberg, D.: Genetic Algorithms in search, optimization, and machine learning. Addison-Wesley, Reading (1989)
64. Górski, J., Zagórski, M.: Reasoning about trust in IT infrastructures. In: Kolowrocki, K. (ed.) European Conference on Safety and Reliability – ESREL 2005, Balkema, Gdynia-Sopot-Gdansk, Poland, vol. 1, pp. 689–695 (2005)
65. Guth, M.A.S.: A probabilistic foundation for vagueness and imprecision in fault-tree analysis. IEEE Transaction on Reliability 40(5), 563–571 (1991)
66. Hagan, M., Demuth, H., Beale, M.: Neural Network Design. PWS Publishing, Boston (1996)
67. Hagan, M., Menhaj, M.: Training feed-forward networks with the Marquardt algorithm. IEEE Transactions on Neural Networks 5(6), 989–993 (1994)
68. Hall, J.W.: Uncertainty-based sensitivity indices for imprecise probability distributions. Reliability Engineering and System Safety 91(10-11), 1443–1451 (2006)
69. Hehenberger, P., Naderer, R., Zeman, K.: Entwicklungsmethodik und Produktmodelle zur Handhabung von Komplexität und Lösungsvielfalt mechatronischer Systeme wärhend des Produktlebenszyklus - Design Methodology and Product Models for the Handling of Complexity and Solution Variety of Mechatronic Systems during Product Life Cycle. In: VDI/VDE (ed.) Mechatronik 2005, vol. 2, pp. 933–948. VDI Verlag, Wiesloch (2005)
70. Helton, J., Johnson, J., Oberkampf, W.: Sensitivity Analysis in Conjunction with Evidence Theory Representations of Epistemic Uncertainty. In: 4th International Conference on Sensitivity Analysis of Model Output (SAMO 2004), Santa Fe, Germany (2004)

71. Helton, J.C., Oberkampf, W.L.: Alternative representations of epistemic uncertainty. Reliability Engineering and System Safety 85(1-3), 1–10 (2004)
72. Hiller, M., Müller, J., Roll, U., Schneider, M., Schröter, D., Torlo, M., Ward, D.: Design and Realization of the Anthropomorphically Legged and Wheeled Duisburg Robot ALDURO - Keynote lecture. In: Proceedings of the Tenth World Congress on the Theory of Machines and Mechanisms, Oulu, Finland, vol. 1, pp. 11–17 (1999)
73. Hitziger, T., Bertsche, B.: An approach to determine uncertainties of prior information – the transformation factor. In: Kolowrocki, K. (ed.) European Conference on Safety and Reliability – ESREL 2005, Balkema, Gdynia-Sopot-Gdansk, Poland, vol. 1, pp. 843–849 (2005)
74. Hoffman, F.O., Hammonds, J.S.: Propagation of Uncertainty in Risk Assessments: The Need to Distinguish Between Uncertainty Due to Lack of Knowledge and Uncertainty Due to Variability. Risk Analysis 14(5), 707–713 (1994)
75. Hohm, T., Limbourg, P., Hoffmann, D.: A Multiobjective Evolutionary Method for the Design of Peptidic Mimotopes. Journal of Computational Biology 13(1), 113–125 (2006)
76. Holland, J.H.: Adaptation in natural and artificial systems. The University of Michigan Press, Ann Arbor (1975)
77. Hughes, E.J.: Constraint Handling with Uncertain and Noisy Multi-objective Evolution. In: IEEE Congress on Evolutionary Computation (CEC 2001), vol. 2, pp. 963–970. IEEE Press, Seoul (2001)
78. IEC International Electrotechnical Commission: IEC 61508 Functional Safety of electrical/ electronic/ programmable electronic safety-related systems. IEC, Geneva, Switzerland (2001)
79. IEC International Electrotechnical Commission: IEC 60601-1 Medical electrical equipment - Part 1: General requirements for basic safety and essential performance. IEC, Geneva, Switzerland (2005)
80. Isermann, R.: Fehlertolerante mechatronische Systeme. In: VDI/VDE (ed.) Mechatronik 2005, vol. 1, pp. 35–68. VDI Verlag, Wiesloch (2005)
81. Jäger, P., Bertsche, B.: An approach for early reliability evaluation of mechatronic systems. In: Kolowrocki, K. (ed.) European Conference on Safety and Reliability – ESREL 2005, Balkema, Gdynia-Sopot-Gdansk, Poland, vol. 1, pp. 925–932 (2005)
82. Jäger, P., Bertsche, B.: Zuverlässigkeitsanalyse mechatronischer Systeme in frühen Entwicklungsphasen - Einordnung in das V-Modell (in German). DVM-Bericht 901, 73–82 (2006)
83. Kerscher, W., Booker, J., Meyer, M., Smith, R.: PREDICT: a case study, using fuzzy logic. In: Reliability and Maintainability, 2003 Annual Symposium - RAMS 2003, Tampa, USA, pp. 188–195 (2003)
84. Klir, G.J.: Facets of Systems Science. Plenum Press, New York (1991)
85. Klir, G.J.: Generalized information theory: aims, results, and open problems. Reliability Engineering and System Safety 85(1-3), 21–38 (2004)
86. Klir, G.J.: Uncertainty and Information: Foundations of Generalized Information Theory. Wiley, New York (2005)
87. Klir, G.J.: Personal communication (2007)
88. Kochs, H.D.: Zuverlässigkeit elektrotechnischer Anlagen (in German). Springer, Berlin (1984)
89. Kochs, H.D.: Key Factors of Dependability of Mechatronic Units: Mechatronic Dependability. In: Panel Session on Risk Management and Dependability - What are the Key Factors? 28th Annual International Computer Software and Application Conference (COMPSAC 2004), pp. 584–586. IEEE Computer Society Press, Hong Kong (2004)
90. Kochs, H.D., Petersen, J.: Verlässlichkeit mechatronischer Systeme (in German). Mitteilungen Fachgruppe fehlertolerierende Rechensysteme 30 (2002)

91. Kochs, H.D., Petersen, J.: A Framework for Dependability Evaluation of Mechatronic Units. In: Brinkschulte, U., Becker, J., Fey, D., Grosspietsch, K.E., Hochberger, C., Maehle, E., Runkler, T. (eds.) ARCS 2004 Organic and Pervasive Computing, Lecture Notes in Informatics (LNI) Proceedings, pp. 92–105. GI-Edition, Augsburg, Germany (2004)
92. Kraftfahrt-Bundesamt: Jahresbericht 2006 (in German). Kraftfahrt-Bundesamt, Flensburg, Germany (2007)
93. Kriegler, E.: Imprecise Probability Analysis for Integrated Assessment of Climate Change. Phd thesis, Universität Potsdam, Potsdam, Germany (2005)
94. Krolo, A., Bertsche, B.: An approach for the advanced planning of a reliability demonstration test based on a Bayes procedure. In: Reliability and Maintainability, 2003 Annual Symposium - RAMS 2003, Tampa, USA, pp. 288–294 (2003)
95. Krolo, A., Rzepka, B., Bertsche, B.: Considering Prior Information for Accelerated Tests with a Lifetime-Ratio. In: 3rd International Conference on Mathematical Methods in Reliability: Methodology and Practice MMR 2002, Trondheim, Norway (2002)
96. Kulturel-Konak, S., Smith, A., Coit, D.: Efficiently Solving the Redundancy Allocation Problem Using Tabu Search. IIE Transactions 35(6), 515–526 (2003)
97. Lambert, H.: Use of Fault Tree Analysis for Automotive Reliability and Safety Analysis. In: 2004 SAE Society of Automotive Engineers World Congress, Detroit, USA (2004)
98. Laprie, J.C.: Dependability: basic concepts and terminology in English, French, German, Italian and Japanese. In: Dependable Computing and Fault-Tolerant Systems, vol. 5. Springer, Wien (1992)
99. Laprie, J.C.: Dependability - Its Attributes, Impairments and Means. In: Randell, B., Laprie, J.C., Kopetz, H., Littlewood, B. (eds.) Predictably Dependable Computing Systems. Springer, New York (1995)
100. Levitin, G.: Optimal series-parallel topology of multi-state system with two failure modes. Reliability Engineering and System Safety 77(1), 93–107 (2002)
101. Levitin, G., Dai, Y., Xie, M., Leng Poh, K.: Optimizing survivability of multi-state systems with multi-level protection by multi-processor genetic algorithm. Reliability Engineering and System Safety 82(1), 93–104 (2003)
102. Limbourg, P.: Multi-objective Optimization of Problems with Epistemic Uncertainty. In: Coello Coello, C.A., Hernández Aguirre, A., Zitzler, E. (eds.) EMO 2005. LNCS, vol. 3410, pp. 413–427. Springer, Heidelberg (2005)
103. Limbourg, P.: Imprecise probability toolbox for MATLAB (2007), `http://www.uni-duisburg-essen.de/il/software.php`
104. Limbourg, P., Germann, D.: Reliability Assessment and Optimization under Uncertainty in the Dempster-Shafer Framework. In: Bedford, T., Cojazzi, G. (eds.) 27th ESReDA Seminar, Glasgow, United Kingdom, pp. 77–89 (2004)
105. Limbourg, P., Germann, D., Petersen, J., Kochs, H.D.: Integration von Verlässlichkeitsbetrachtungen in die VDI 2206: Entwicklungsmethodik für mechatronische Systeme – Ein Anwendungsbeispiel (in German). In: VDI/VDE (ed.) Mechatronik 2005, vol. 2, pp. 899–918. VDI Verlag, Wiesloch (2005)
106. Limbourg, P., Kochs, H.D.: Maintenance scheduling by variable dimension evolutionary algorithms. In: Kolowrocki, K. (ed.) European Conference on Safety and Reliability – ESREL 2005, Balkema, Gdynia-Sopot-Gdansk, Poland, vol. 2, pp. 1267–1275 (2005)
107. Limbourg, P., Kochs, H.D.: Predicting Imprecise Failure Rates from Similar Components: a Case Study using Neural Networks and Gaussian Processes. In: ARCS Workshop Dependability and Fault Tolerance, GI, Frankfurt, Germany, pp. 26–35 (2006)
108. Limbourg, P., Kochs, H.D.: Multi-objective optimization of generalized reliability design problems using feature models – A concept for early design stages. Reliability Engineering and System Safety 93(6), 815–828 (2008)

109. Limbourg, P., Kochs, H.D., Echtle, K., Eusgeld, I.: Reliability Prediction in Systems with Correlated Component Failures - An Approach Using Copulas. In: ARCS Workshop Dependability and Fault Tolerance, 2007, pp. 55–62. VDE-Verlag, Zürich (2007)
110. Limbourg, P., Salazar Aponte, D.E.: An Optimization Algorithm for Imprecise Multi-Objective Problem Functions. In: IEEE Congress on Evolutionary Computation (CEC 2005), vol. 1, pp. 459–466. IEEE Press, Edinburgh (2005)
111. Limbourg, P., Savić, R., Petersen, J., Kochs, H.D.: Approximating Failure Distributions from Similar Components Using Artificial Neural Networks. In: Soares, C.G., Zio, E. (eds.) European Conference on Safety and Reliability – ESREL 2006, Estoril, Portugal, pp. 911–919 (2006)
112. Limbourg, P., Savić, R., Petersen, J., Kochs, H.D.: Fault Tree Analysis in an Early Design Stage using the Dempster-Shafer Theory of Evidence. In: Aven, T., Vinnem, J.E. (eds.) European Conference on Safety and Reliability – ESREL 2007, vol. 2, pp. 713–722. Taylor & Francis, Stavanger (2007)
113. Lindskog, F.: Modelling Dependence with Copulas and Applications to Risk Management. Master thesis, Eidgenössische Technische Hochschule Zürich, Zürich, Switzerland (2000)
114. Litto, M., Lewek, J.: Interdisziplinäres, funktionales Engineering konventioneller Maschinen und Anlagen (in German). In: Gausemeier, J., Rammig, F.J., Schäfer, F., Wallaschek, J. (eds.) 3. Paderborner Workshop Intelligente mechatronische Systeme, vol. 163, pp. 101–112. HNI-Verlagsschriftenreihe, Paderborn (2005)
115. MacKay, D.J.C.: Gaussian processes - a replacement for supervised neural networks? Lecture notes for a tutorial. In: Advances in Neural Information Processing Systems NIPS. MIT Press, Denver (1997)
116. Mardia, K.V.: Families of Bivariate Distributions. Charles Griffin and Co. Ltd., London (1970)
117. Marseguerra, M., Zio, E.: Basics of the Monte Carlo Method with Application to System Reliability. LiLoLe Publishing, Hagen (2002)
118. Marseguerra, M., Zio, E., Podofillini, L., Coit, D.: Optimal Design of Reliable Network Systems in Presence of Uncertainty. IEEE Transactions on Reliability 54(2), 243–253 (2005)
119. Möhringer, S.: Gibt es ein gemeinsames Vorgehen in der Mechatronik (in German). In: VDI/VDE (ed.) Mechatronik 2005, vol. 1, pp. 229–252. VDI Verlag, Wiesloch (2005)
120. MoD Ministry of Defence: Defence Standard 00-42: R&M Case, Reliability and Maintainability (R&M) Assurance Guidance, Defence Procurement Agency, Glasgow, United Kingdom, vol. 3 (2003)
121. Moral, S.: Personal communication (2005)
122. Murphy, N.: Watchdog Timers. Embedded Systems Programming, 79–80 (October 2001)
123. Nelsen, R.B.: An Introduction to Copulas. Springer, New York (1999)
124. Nilsen, T., Aven, T.: Models and model uncertainty in the context of risk analysis. Reliability Engineering and System Safety 79(3), 309–317 (2003)
125. Oberkampf, W.L., Helton, J.C., Joslyn, C.A., Wojtkiewicz, S.F., Ferson, S.: Challenge problems: uncertainty in system response given uncertain parameters. Reliability Engineering and System Safety 85(1-3), 11–19 (2004)
126. Pareto, V.: Manual of political Economy (Reprint). A.M. Kelley, New York (1971)
127. Patë-Cornell, E.: Risk and uncertainty analysis in government safety decisions. Risk Analysis 22(3), 633–646 (2002)
128. Pfeifer, T., Schmidt, R., Geisberger, E.: Integriertes Projektmanagement und Requirements Engineering für die Entwicklung von eingebetteten Systemen. In: Gausemeier, J., Rammig, F.J., Schäfer, F., Wallaschek, J. (eds.) 3. Paderborner Workshop Intelligente mechatronische Systeme, vol. 163, pp. 85–100. HNI-Verlagsschriftenreihe, Paderborn (2005)
129. Pham, H.: Reliability Engineering. Springer, London (2003)

130. Pozsgai, P., Krolo, A., Bertsche, B., Fritz, A.: SYSLEB-a tool for the calculation of the system reliability from raw failure data. In: Reliability and Maintainability, 2002 Annual Symposium - RAMS 2002, La Jolla, USA, pp. 542–549 (2002)
131. Prampolini, A.: Modelling default correlation with multivariate intensity processes (2001), http://www.math.ethz.ch/~degiorgi/CreditRisk/CR1.html
132. Raich, A.M., Ghaboussi, J.: Implicit representation in genetic algorithms using redundancy. Evolutionary Computation 5(3), 277–302 (1997)
133. Rakowsky, U.K.: Some Notes on Probabilities and Non-Probabilistic Reliability Measures. In: Kolowrocki, K. (ed.) European Conference on Safety and Reliability – ESREL 2005, Balkema, Gdynia-Sopot-Gdansk, Poland, vol. 2, pp. 1645–1654 (2005)
134. Rakowsky, U.K., Gocht, U.: Modelling of Uncertainties in Reliability Centred Maintenance – A Dempster-Shafer Approach. In: Aven, T., Vinnem, J.E. (eds.) European Conference on Safety and Reliability – ESREL 2007, vol. 1, pp. 739–745. Taylor & Francis, Stavanger (2007)
135. Ramakrishnan, N., Bailey-Kellogg, C., Tadepalli, S., Pandey, V.: Gaussian Processes for Active Data Mining of Spatial Aggregates. In: Proc. SIAM Data Mining Conference. SIAM, Newport Beach, USA (2005)
136. Rechenberg, I.: Evolutionsstrategie - Optimierung technischer Systeme nach Prinzipien der biologischen Evolution (in German). Frommann-Holzboog, Stuttgart (1973)
137. ReliaSoft: Characteristics of the Weibull Distribution (2002), http://www.weibull.com/hotwire/issue14/index.htm
138. Rensselaer Polytechnic Institute: Mechatronics (2007), http://mechatronics.rpi.edu
139. Rocquigny, E.: A statistical approach to control conservatism of robust uncertainty propagation methods; application to accidental thermal hydraulics calculations. In: Kolowrocki, K. (ed.) European Conference on Safety and Reliability – ESREL 2005, Balkema, Gdynia-Sopot-Gdansk, Poland, vol. 2, pp. 1691–1699 (2005)
140. Sallak, M., Simon, C., Aubry, J.F.: Evaluating Safety Integrity Level in presence of uncertainty. In: 4th International Conference on Safety and Reliability, KONBIN 2006, Kraków, Poland (2006)
141. Saltelli, A., Tarantola, S., Campolongo, F., Ratto, M.: Sensitivity Analysis in Practice - A Guide to Assessing Scientific Models. Wiley, New York (2004)
142. Sandtorv, H.A., Østebø, R., Kortner, H.: Collection of reliability and maintenance data - development of an international standard. In: Kolowrocki, K. (ed.) European Conference on Safety and Reliability – ESREL 2005, Balkema, Gdynia-Sopot-Gdansk, Poland, vol. 2, pp. 1751–1756 (2005)
143. Savić, R.: Neural generation of uncertainty reliability functions bounded by belief and plausibility frontiers. In: Kolowrocki, K. (ed.) European Conference on Safety and Reliability – ESREL 2005, Balkema, Gdynia-Sopot-Gdansk, Poland, vol. 2, pp. 1757–1762 (2005)
144. Savić, R., Limbourg, P.: Aggregating Uncertain Sensor Information for Safety Related Systems. In: Guedes Soares, C., Zio, E. (eds.) European Conference on Safety and Reliability – ESREL 2006, Estoril, Portugal, pp. 1909–1913 (2006)
145. Schölkopf, B., Smola, A.J.: Learning with Kernels: Support Vector Machines, Regularization, Optimization, and Beyond. MIT Press, Cambridge (2002)
146. Schneeweiss, W.G.: Boolean functions with engineering applications and computer programs. Springer, New York (1989)
147. Schoo, H., Schubert, R.: Qualitäts- und Risikomanagement von Entwicklungsprozessen (in German). In: risk.tech 2004. TÜV Automotive GmbH, TÜV SÜD Akademie GmbH, München, Germany (2007)
148. Schwefel, H.P.: Projekt MHD-Staustrahlrohr: Experimentelle Optimierung einer Zweiphasendüse, Teil I (in German). Technischer Bericht 11.034/68, 35. AEG Forschungsinstitut, Berlin, Germany (1968)

149. Sentz, K., Ferson, S.: Combination of Evidence in Dempster-Shafer Theory. No. SAND2002-0835 in Sandia Report. Sandia National Laboratories, Albuquerque, USA (2002)
150. Shafer, G.: A Mathematical Theory of Evidence. Princeton University Press, Princeton (1976)
151. Shafer, G., Logan, R.: Implementing Dempster's rule for hierarchial evidence. Artificial Intelligence 33(3), 271–298 (1987)
152. Shannon, C.: A mathematical theory of communication. The Bell System Technical Journal 27, 379–423 (1948)
153. Silvey, S.D.: Statistical Inference. In: Monographs on Statistics and Applied Probability. Chapman and Hall, London (1975)
154. Sklar, A.: Fonctions de répartition á n dimensions et leurs marges. Publications. de l'Institut de statistique de l'Université de Paris 8, 229–231 (1959)
155. Smets, P.: Imperfect information: Imprecision - Uncertainty. In: Motro, A., Smets, P. (eds.) Uncertainty Management in Information Systems. From Needs to Solutions. Kluwer, Dordrecht (1997)
156. Smith, D., Simpson, K.: Functional Safety - A Straightforward Guide to Applying IEC 61508 and Related Standards. Elsevier, Oxford, United Kingdom (2004)
157. Soundappan, P., Nikolaidis, E., Haftka, R.T., Grandhi, R., Canfield, R.: Comparison of evidence theory and Bayesian theory for uncertainty modeling. Reliability Engineering and System Safety 85(1-3), 295–311 (2004)
158. Srinivas, N., Deb, K.: Multiobjective function optimization using nondominated sorting genetic algorithms. Evolutionary Computation 2(3), 221–248 (1995)
159. Teich, J.: Pareto-Front Exploration with Uncertain Objectives. In: Zitzler, E., Deb, K., Thiele, L., Coello Coello, C.A., Corne, D.W. (eds.) EMO 2001. LNCS, vol. 1993, pp. 314–328. Springer, Heidelberg (2001)
160. The MathWorks: MATLAB Neural Network Toolbox (2007)
161. Tian, Z., Zuo, M.J., Huang, H.: Reliability-redundancy allocation for multi-state series-parallel systems. In: Kolowrocki, K. (ed.) European Conference on Safety and Reliability – ESREL 2005, Balkema, Gdynia-Sopot-Gdansk, Poland, vol. 2, pp. 1925–1930 (2005)
162. Tonon, F.: Using random set theory to propagate epistemic uncertainty through a mechanical system. Reliability Engineering and System Safety 85(1-3), 169–181 (2004)
163. US Department of Defense: MIL-HDBK-217F Reliability Predition of Electronic Equipment. Department of Defense, Washington DC, USA (1991)
164. Utkin, L.V.: A second-order uncertainty model for calculation of the interval system reliability. Reliability Engineering and System Safety 79(3), 341–351 (2003)
165. Utkin, L.V.: Reliability models of m-out-of-n systems under incomplete information. Computers and Operations Research 31(10), 1681–1702 (2004)
166. Utkin, L.V., Coolen, F.P.A.: Imprecise reliability: an introductory overview. In: Levitin, G. (ed.) Computational Intelligence in Reliability Engineering, vol. 2, pp. 261–306. Springer, New York (2007)
167. Utkin, L.V., Kozine, I.O.: Computing System Reliability Given Interval-Valued Characteristics of the Components. Reliable Computing 11(1), 19–34 (2005)
168. VDA Verband der Automobilindustrie: Zuverlässigkeitssicherung bei Automobilherstellern und Lieferanten, Teil 1: Zuverlässigkeitsmanagement (in German). Qualitätsmanagement in der Automobilindustrie. VDA, Frankfurt, Germany (2000)
169. VDI Verein Deutscher Ingenieure: VDI 2206 Entwicklungsmethodik für mechatronische Systeme - Design methodology for mechatronic systems. Beuth, Berlin, Germany (2004)
170. Walley, P.: Statistical Reasoning with Imprecise Probabilities. Chapman and Hall, London (1991)

171. Walter, M., Schneeweiss, W.: The modeling world of reliability/safety engineering. LiLoLe Publishing, Hagen (2005)
172. Walter, M., Trinitis, C., Karl, W.: Evaluierung von verteilten hochverfügbaren Systemen mit OpenSESAME (in German). In: PARS 2001 Workshop, Parallel-Algorithmen und Rechnerstrukturen, Gesellschaft für Informatik e.V., München, Germany, pp. 29–39 (2001)
173. Weston, J., Elisseeff, A., Bakir, G., Sinz, F.: Spider: object-orientated machine learning library (2007),
`http://www.kyb.tuebingen.mpg.de/bs/people/spider/`
174. WWWBar: Weibull Reliability Database For Failure Data For Various Components (2007), `http://www.barringer1.com/wdbase.htm`
175. WWWCondProb: Conditional probability (2007),
`http://www.uvm.edu/~dhowell/StatPages/Glossaries/GlossaryI-L.html`
176. WWWIndProb: Statistical independence (2007),
`http://en.wikipedia.org/wiki/Statistical_independence`
177. WWWJointProb: Glossary: Joint probability (2007),
`http://www.uvm.edu/~dhowell/StatPages/Glossaries/GlossaryI-L.html`
178. WWWSandia: Epistemic Uncertainty Project Home (2007),
`http://www.sandia.gov/epistemic/`
179. WWWSip: SIPTA Home Page (2007),
`http://www.sipta.org/`
180. Yager, R.R.: Arithmetic and other operations on Dempster-Shafer structures. International Journal of Man-Machine Studies 25, 357–366 (1986)
181. Yager, R.R.: On the Dempster-Shafer framework and new combination rules. Information Sciences 41, 93–137 (1987)
182. Zadeh, L.: Fuzzy sets as a basis for a theory of possibility. Fuzzy Sets Syst. 1, 3–28 (1978)
183. ZF Friedrichshafen AG: ZF ASTronic (2007),
`http://www.zf.com/`
184. Zitzler, E., Laumanns, M., Thiele, L.: SPEA2: Improving the Strength Pareto Evolutionary Algorithm. In: Giannakoglou, K.C., Tsahalis, D.T., Periaux, J., Papaillou, K.D., Fogarty, T. (eds.) EUROGEN 2001 - Evolutionary Methods for Design, Optimisation and Control with Applications to Industrial Problems, Barcelona, Spain (2001)

Index